Active Radar
Electronic
Countermeasures

Edward J. Chrzanowski

The Artech House Radar Library

David K. Barton, *Series Editor*

Active Radar
Electronic
Countermeasures

Edward J. Chrzanowski

Artech House

Library of Congress Cataloging-in-Publication Data

Chrzanowski, Edward, J., 1919-
  Active radar electronic countermeasures / Edward J. Chrzanowski.
      p.   cm.
    ISBN 0-89006-290-0
    1. Radar--Military applications. 2. Radar--Interference.
3. Electronic countermeasures. 4. Radar transmitters. I. Title.
UG612.C47   1989   89-17553
623'.7348-dc20   CIP

ARTECH HOUSE, INC.
685 Canton Street
Norwood, MA 02062

International Standard Book Number: 0-89006-290-0
Library of Congress Catalog Card Number: 89-17553

10   9   8   7   6   5   4   3   2   1

This book is dedicated to my two grandchildren,
Stephan and Christopher Khanoyan

Contents

Preface

This book was written to support an intensive short course on *Active Electronic Countermeasures* sponsored by the Technology Service Corporation. The material in the course is presented as a subset of electronic warfare and is concerned primarily with *electronic countermeasures* (ECM) systems which generate and radiate signals to interfere with the proper operation of hostile radar systems. The course is designed for those with a generalized technical background, and includes a review of the various classes of radar systems and the ECM signal environment. The course was presented in open sessions to the general public and in closed sessions at various government and industrial facilities.

The course and this book are concerned exclusively with active electronic countermeasures systems. These are jamming systems which generate and radiate signals at or near the radar carrier frequency and are properly modulated to interfere with the radar's ability to detect and measure the position parameters of the radar target. An important advantage of these systems, when compared to passive countermeasures such as chaff and corner reflectors, is that modulations in time, frequency or amplitude can be imposed on the interfering signals as required for maximum effectiveness against more advanced radars.

Although other references were available to support the material in the course, none dealt exclusively with actively radiating systems designed to jam the radars used in defense systems. All the references available were concerned with the broad class of electronic warfare, with less than adequate attention to the important aspect of radar system jamming.

Because of the author's experience in radar design as a systems engineer at the Westinghouse Defense Center, the conclusions were easily reached that radar is a significant contributor to successful defense system operation, and that interference with its proper operation could seriously jeopardize the defense process. As a radar designer, the author became intimately familiar with vulnerabilities in radar which could be exploited by appropriate ECM.

Because Westinghouse has been a leader in the development and production of defense system radars (both surveillance and fire control, airborne as well as surface based), ECM system designers at that facility were provided the opportunity to consult with radar designers as to the effectiveness of their proposed jamming techniques. In addition, the radar designers were able to consult the ECM designers as to the types of jamming techniques being developed, and to determine the effectiveness of any possible counters to those techniques being incorporated in the radars.

The Vietnam conflict added testimony to the importance of radar in the guidance of missiles to intercept aircraft penetrating into enemy territories. During that conflict, aircraft losses rose to such a level that countermeasures to proper operation of radars were eagerly sought by the armed forces. This requirement initiated the formation of the Wild Weasel Squadron, the responsibility of which was to jam the defense radars so that the friendly force aircraft could penetrate enemy defenses with relative impunity. In addition, electronic countermeasure systems were developed to be carried by the penetrators to provide self-protection in addition to the mutual support provided by Wild Weasel aircraft. The effectiveness of the countermeasures soon became evident with the deep reduction in aircraft losses when countermeasures were employed against the defense radars.

The author was privileged to serve under the leadership and guidance of Joseph Legin, the manager of the electronic warfare department at Westinghouse during that period. Mr. Legin and his department were asked by the US Air Force to produce systems which would reduce the effectiveness of the adversary's radar systems. Under a Quick Reaction Capability (QRC) program, and with Mr. Legin's personal committment to the Wild Weasel pilots, the department delivered systems within five months of award of contract.

Because of the lack of documentation and references concerning this relatively new concept, many innovative jamming techniques and associated system analyses were needed to support that development. This book is a compilation of that work.

The first chapter in the book provides a general overview of the types of radars which can be expected as a vehicle enters hostile territories. A general overview of the target position measurement techniques used by radars is given, in addition to suggestions as to possible techniques which may be used to interfere with the radar processes.

Because of the importance of generating an interfering signal that must be of adequate power to compete with the true target return, the second chapter provides a detailed discussion on the determination of the *jamming-to-signal ratio* (*J/S*) as well as the *effective radiated power* (ERP) required by a jammer to produce the required *J/S*. The difference between a constant-power and a constant-gain ECM system is also discussed.

Chapters 3 through 6 describe ECM techniques developed and used against the various search and track radars expected in the threat environment. The format used provides one chapter each for the range and angle measurement functions of surveillance radars, and one chapter each for the range and angle measurement functions of the fire control or tracking radars. Although some of the techniques discussed have become obsolete because of advances made in radar designs, they are included for the sake of completeness as well as to indicate how and why advances in radar countermeasures evolved.

Chapter 7 discusses other radars which do not necessarily fall into the category of either search of track radars, but are significant to defense systems.

Chapter 8 discusses the requirements for a receiver and data processing system properly managing the modulations and power of the ECM system for simultaneous and effective jamming of different radars in a heavy radar threat environment. This function of the ECM system is currently receiving the most interest and funds in the industry because of the requirement for the system to operate in a very dense and flexible radar environment.

Chapter 9 closes the discussion of this fascinating subject with an outline of advancements being considered in future radar systems to improve the capabilities of the radar when confronted by modern ECM systems. As we indicate, the confrontation between radar designers and ECM designers is never-ending, with no definite winner ever emerging. Because of the effectiveness of ECM systems, defense systems have resorted to the use of other sensing devices that use infrared and laser techniques.

The author expresses appreciation to Mr. Gene Fox, formerly the manager of the engineering section of the electronic warfare department at Westinghouse, for the time and effort that he contributed in editing the manuscript and offering several important suggestions to improve the quality of the book.

Chapter 1
ECM Overview

1.1 INTRODUCTION

1.1.1 History

At one time, vehicles penetrating enemy territories were able to approach under cover of darkness, rain, fog, and even man-made smoke screens to conceal their presence from their adversaries. The development of radar rendered all of these options ineffective, forcing penetration vehicles to seek other means of concealing their approach. One method employed, still very effective, was to make the approach at very low altitudes and with very low detection profiles to take advantage of the radar's inability to "see" over the horizon.

Another method pursued that still receives much attention is the reduction of the echo power reflected from the protected vehicle. This is achieved by reducing the effective radar reflective area of the vehicle, using special design techniques and absorptive coatings on the vehicle's reflecting surface; this approach is commonly known as the *stealth* technique.

There is still considerable interest in both of these methods, but they have serious limitations. The low altitude approach requires sophisticated navigation systems to allow the penetrating vehicle to follow the contour of the earth's surface to reach designated targets. Furthermore, because aircraft weapon delivery is more effective when launched from higher altitudes, the aircraft are required to abandon the low altitude approach over the target area. The value of the stealth technique depends on the adversary's ability to redesign his radars to compensate for the loss in radar cross section to maintain the same detection range. The radar design would need to include improvements in transmitter power, antenna gain, or receiver sensitivity. For each dB of reduction in radar cross section, a corresponding increase in any one or in all of these is required to maintain the same radar detection range. For example, a 20 dB reduction will cause a reduction in radar detection range to one-third of its original capability.

A third method of concealing the presence of penetrating vehicles from hostile radars is with the use of active electronic countermeasures, which is the subject of this book. Active ECM systems are designed to cause the injection of signals into the radar's receiving circuits, which interfere with the true reflected signals in a manner that denies the radar's ability to detect or locate the vehicles being protected by the ECM. Although radar reflective devices such as chaff and corner reflectors have been and are effectively used to inject such interfering signals into a radar, this book deals exclusively with devices which deliberately generate and radiate signals for the purpose of competing directly with the echo signal entering the radar receiver.

All three methods of protecting a vehicle penetrating hostile territory—stealth, low altitude and ECM—can be used individually or in any combination The design of a system using one method must be in accordance with its usage with either of the other two methods. With a stealth aircraft, for example, the radiation from the ECM system must not betray the otherwise undetectable presence of the vehicle.

Although the use of communications provides a very important function in a defense system, that subject as well as jamming against communications are not directly addressed in this book. Excellent discussions on that subject are presented in other references. This book is primarily concerned with jamming systems which interfere with radar receivers.

Because of the sensitive nature of ECM, written or printed material on the subject was not always readily available to interested readers who did not possess the proper security clearances. Engineers committed to designing and evaluating the performance of these systems were required to apply and extrapolate on knowledge obtained in the study of radars and other related disciplines to carry out their responsibilities. The author of this book was among those who were required to develop new theories, formulas and concepts to advance the progress of this relatively new area of study.

Although crude forms of jamming, such as dispensed chaff, and unsophisticated forms of noise jamming were employed during World War II, it was not until radar designers applied antijamming (AJ) techniques into their systems that more sophisticated jamming techniques were needed. This demanded new techniques and technology to counter the AJ techniques, as well as the development of new theories and concepts to evaluate their performance. The net result was and is a continuing battle between radar designers and ECM designers. This aspect of electronic warfare makes the subject of active electronic countermeasures so interesting and fascinating. The objective of this book is to explain how the current technology of ECM has evolved as a result of this continuing exercise between ECM and radar designers, and to prepare the interested student for future requirements.

Several books have appeared in recent years which address the subject of ECM. These generally include jamming techniques applicable to communications systems and optical sensing systems. They also include passive types of countermeasures, such as chaff, flares, and corner reflectors. Although these books include discussions of active ECM, the material pertaining to the subject is interleaved with the other discussions.

This book discusses jamming techniques used against radar systems of all varieties, namely noncoherent pulsed, coherent pulsed and FMCW doppler. Furthermore, this book restricts its discussions on jamming systems to those which deliberately radiate signals for the purpose of interfering with a victim radar. The modulations of the interfering signals are designed to deny the radar its ability to detect, identify, or locate targets to the degree required for effective performance of the defense system. Passive devices, such as chaff and corner reflectors, are only discussed in passing in this book.

1.1.2 Component Requirements

Because of the unique requirements of ECM systems, many new components were needed and developed; these included the traveling-wave tube, the microwave signal memory system, broad frequency band antennas, attenuators, and phase shifters. These developments included new techniques of ECM system design as well as new techniques for evaluation of performance. Furthermore, because of the speed of response required in very dense radar environments, high speed logic circuits and extremely short delay microwave amplifying systems were needed. In addition, systems had to be designed to transmit microwave signals which, upon arriving at the victim radar's receiving antenna, could not be distinguished from true reflected signals from targets in the environment. The latter requirement became formidable with the advent of complex intrapulse modulations of the radar signal.

Earlier ECM systems operated on the basis of primitive information as to the radar environment which threatened the protected vehicle during its penetration into hostile territories. The jamming configuration and its use were almost always based on *a priori* data as to the expected environment during any one mission. In that case, two possibilities existed: that the jamming may have only minimum effectiveness against the radars of interest, and that friendly radars in the environment may be jammed.

Additionally, indiscriminate employment of jamming systems could provide premature radar detection of the presence of penetrating vehicles. This is due to the fact that the jamming could result in signal levels into the radar detection circuits that will be larger than the radar reflection from the protected vehicle. Furthermore, the mode of operation of the ECM system and the signal modulation parameters were not determined on the basis of the existing threat environment. In too many cases, the power transmitted by the ECM system was at radar frequencies that did not include the frequency of the intended victim radar, which was an obvious waste of valuable energy.

1.1.3 ECM Receiver Requirements

Because of the numerous advances made in radar design (many were required to counter ECM devices), customizing the modulations imposed on the ECM transmitted signal became necessary so that it would be effective against certain types of radars. In addition, because of the proliferation of radars in the environment and increases in their transmitter power levels, directing ECM-generated energy against those radars which imposed the greatest threat to the survival of the vehicle being protected became necessary. This requirement necessitated the capability of not only detecting the presence of hostile radar radiations, but also identifying the radar as to its type and intent. In addition, determining the radar's location became necessary, at least in angle, to aim ECM-transmitted energy in the direction of the most serious threats.

Because of the requirement for rapid response to potential threats in the environment, making these determinations in as short a time as possible became necessary. Ideally, it would be desirable to respond effectively at the first instant that a signal from a radar associated with a threat system is detected. However, reliable detection and analysis of intercepted signals dictate a measurable period of time for effective operation. This requirement is aggravated by the fact that many other signals exist in the environment that are of no threat to the penetrating vehicle or of no interest to the ECM system.

Nevertheless, these signals enter the ECM receiver and must be separated from those of interest. Too often, the filtering process required to reject signals of no interest is identical to that required to accept those of interest, and these signals only serve to saturate the receiver processing circuits. The net result is a requirement for high volume and very high speed processing circuits, said requirement being satisfied with the development of *microwave monolithic integrated circuits* (MMICs) and *very high speed integrated circuits* (VHSICs).

The ability to direct ECM-transmitted energy toward radars of interest has dictated the requirement for high speed, steerable antennas. This requirement is being satisfied with the development of very agile, electronically steerable antenna arrays; beam switching speeds to within a fraction of a microsecond are being realized with current systems.

Furthermore, because of the proliferation of defense radar systems as well as the influx of new radar techniques to counter the ECM devices being developed by jammer designers, greater selectivity has become necessary in determining which of the intercepted radar signals is a more immediate threat to the survivability of the penetrating vehicle, as well as the type of modulations which are most effective against each such threat.

In most cases, individual threat radars require different types of modulations, even though simultaneous operation must be used against them. Modern ECM systems are being developed that impose modulations on transmitted signals on a pulse-to-pulse or system time-shared basis. This requires the ability to anticipate the arrival of each of the pulses from any one radar and to generate the type of modulation which must be imposed on the pulse or pulses as they pass through the transmitter system.

1.1.4 System Requirements

As we will show in the next chapter, the power level of the radar return reflected off the target of interest is directly proportional to the radar cross-sectional area of the vehicle; this may or may not be the physical cross-sectional area of the object, depending on the design of the structure. Although much effort has been and is being expended to reduce the reflective area by using absorptive coatings and judicious design of the structure, reductions of 20 to 30 dB have been difficult to achieve. Unfortunately for vehicle designers, the value of such reductions must be weighed against the enemy's ability to compensate for the effect of the reduction in the design of the radar. Increasing the radar transmitter power, its antenna gain or its receiver sensitivity by an equal amount to recover its original detectability negates the protection gained with the stealth technique.

We assume throughout this book that a target signal return from the protected vehicle exists at the radar antenna terminals. Whether the signal is detectable by the radar depends on the level of signal, the sensitivity of the radar receiver, and the level of interference accompanying the signal in the detection circuits.

This suggests the real possibility that ECM action can prematurely alert radar defense systems by transmitting a stronger signal than is reflected from the target, even when that signal may not have been detectable by the radar. Appendix B addresses this subject with calculations that depict required radar and ECM receiver characteristics to ensure that the radar detects the target before the target can detect the radar signal. In general, because of the excellent sensitivities attributable to current ECM receivers, we can expect that the ECM receiver will often detect the radar signal before the radar can detect the target skin return. As shown in the appendix, unless the radar controls its radiated power to yield target detections only to low radar ranges, the ECM will almost always detect the radar first, even with receiver sensitivities as low as -60 dBm.

To counter this advantage of the ECM receiver, radar designers have incorporated sophisticated intrapulse modulations in the more advanced radars. By using appropriate coding of the modulations, the radar enjoys a signal "processing gain" which is available only to receivers that contain the processing code. Generally, this code cannot be reliably measured at the ECM receiver, even though the signal may be detectable. One of the effects of the intrapulse modulations is to spread the spectrum of the radar transmitted signal so that the received signal at the ECM receiver is below the noise level of that receiver. Radars which employ this technique are known as *low probability of intercept* (LPI) radars. Improvements in the signal-to-noise ratio at the ECM receiver may help to pull the radar signal out of the noise, which will make the signal detectable; however, the ECM receiver, without the code, is still not able to decode the intrapulse modulations, as can the radar receiver.

Because we assume that a target signal is available to the radar at its antenna terminals, jamming systems are required to counter this signal by radiating another signal with sufficient power to enter the radar simultaneously with the target return or, if not simultaneous, in a manner that serves to confuse the radar detection and analysis circuits as to the presence or location of its intended target.

We emphasize that ECM systems do not remove true target returns; they generate signals that enter the radar receiver with a power level sufficient to conceal the true target return by covering it with noise or false returns, or causing it to be lost among many other returns. Indeed, even if the ECM signal conceals the true target returns, it must also contain modulations different from those which exist on the true target return in order to confuse the location measurement circuits of the radar as to the true location of the target or the ECM system.

The task for active ECM then is to generate enough power to override the data on the true signal and, having done so, ensure that the modulations imposed on the ECM transmitted signal are acceptable to the radar analysis circuits, but with erroneous data as to the location of its intended target. Whether the ECM transmitted power level is adequate depends on the modulations used; these are discussed in the following chapters.

1.2 ELECTRONIC WARFARE

Figure 1.1 depicts the three basic aspects of electronic warfare (EW): electronic support measures (ESM), electronic countermeasures (ECM), and electronic counter-countermeasures (ECCM).

1.2.1 Electronic Support Measures (ESM)

Electronic support measures consist of systems which intercept and analyze all radiated signals in the environment to determine the optimum response to be made by a vehicle to improve its probability of survival. These responses can be of any variety (i.e., active or passive expendables, maneuvers, or ECM).

```
                        ELECTRONIC WARFARE
                             (EW)
                              |
        _____
       |                      |                         |
       |                      |                         |
   ELECTRONIC                 |                     ELECTRONIC
 SUPPORT MEASURES             |              COUNTER-COUNTERMEASURES
     (ESM)          ELECTRONIC COUNTERMEASURES          (ECCM)
                             (ECM)
                              |
                       _____
                      |              |
                    ACTIVE        PASSIVE
```

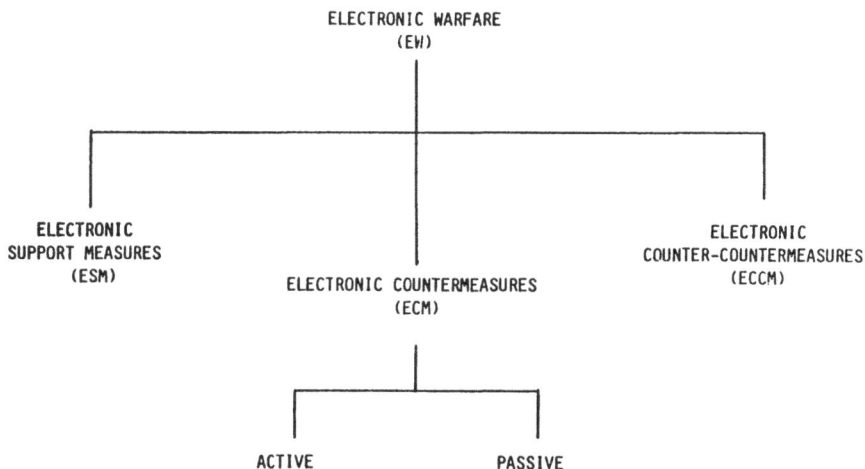

Figure 1.1 Components of electronic warfare.

The responsibility of ESM then is to detect and extract all the information it can from intercepted signals to determine the type of defense system as well as its intent relative to the vehicle the ESM is designed to protect. The bandwidth of the ESM system is most often extremely wide, and may even include infrared (IR) and optical detectors. The bandwidth should be at least as great as any one of its responses, including that of wide-bandwidth ECM systems.

Although ESM is shown here as a separate entity relative to ECM system operation, modern and future ECM systems will contain their own integrated receivers which will provide whatever detection and analysis is required for effective ECM operation. Indeed, the *integrated electronic warfare system* (INEWS) being developed for future aircraft has the capability of intercepting and analyzing signals over very wide radio frequency bands and integrating that data with data obtained from other sensors (radar, IR, optics) on board the aircraft. The data will then be used to determine and activate the optimum response to the threat environment.

1.2.2 Electronic Counter-countermeasures (ECCM)

Another study under the umbrella of electronic warfare is the study of *electronic counter-countermeasures* (ECCM). Generally, this study is undertaken by radar designers to counter devices developed by ECM designers to disrupt the operation of their radars. This results in a continuing conflict between the ECM and radar designers to counter each other's techniques, which is the factor that makes the study of ECM or ECCM fascinating.

1.2.3 Electronic Counter-measures (ECM)

The third area of study in electronic warfare is that of *electronic countermeasures*. As shown in the figure, this study is divided into two principal areas, active and passive electronic countermeasures. *Active electronic countermeasures,* the subject of this book, consists of all devices that deliberately generate and radiate radio frequency signals to compete with true radar reflected signals in order to disrupt the intended function of the victim radar. These include

- Noise Jammers
- Deception Repeaters and Transponders
- Electronic Decoys

Passive electronic countermeasures are those devices which are designed to reflect the intercepted radar radiation in such a manner that the reflections compete with true target returns to conceal the true target position. These include chaff and corner reflectors.

1.3 RADAR ENVIRONMENT

It is not the intent of this book to describe or identify specific known threat radar systems or their characteristics, or to identify or describe friendly or hostile electronic warfare systems. All radars and ECM systems are described in a generic fashion; anyone concerned with actual systems can easily apply specific characteristics if desired without loss of clarity. We show that all radars operate on the basis of a few basic principles; the only difference being in methods of mechanization of the required functions. All radars can be catalogued into one of three basic types: (1) those which primarily search for targets; (2) those which track one or more selected or assigned targets; and (3) those which perform tracking of targets simultaneously with a search mode.

As a vehicle penetrates hostile territory, the following electronic contact can be expected in the order listed as the vehicle enters:

- Early Warning Radars
- Height Finder Radars
- Acquisition Radars
- Fire Control Radars
- Terminal Guidance Radars
- Missile Fuze

1.3.1 Early Warning Radars

As indicated, the first contact would very likely be with an early warning radar (EWR). These can be either ground-based, shipborne, airborne or space-based.

Their primary function is to search a large volume of space in which penetrators can be expected; they perform only limited tracking of targets, if at all. These radars are generally designed with very high powered transmitters and large antennas to provide very long detection ranges, even against vehicles with a small radar cross-sectional area. Because these radars are required to search a large volume, the data rate for any target is inadequate to perform any fine range or angle measurement; only rough measurements of target locations are provided.

These radars are primarily used for alerting or cueing other sensors, which provide the required accuracy for interception of any intruders into the territory. Target detection was generally performed visually by an operator when conventional pulsed radars were used for the EWR function. However, more sophisticated radars are resorting to the use of electronic detection and high speed computer analysis for detection, identification or location of all detectable targets in the environment.

This is especially true in the case of a doppler-type radar which must extract its targets of interest from powerful natural interfering signals. In these cases, the actual signal reflected from the target and detected by the electronic circuits cannot be displayed in its "raw" form to an operator; the targets and their locations are displayed synthetically in a doppler radar. Although these radars present cleaner and clearer targets to the operator, they also deny the operator the chance to "read through" interfering signals as can be done when the "raw video" is presented.

Unfortunately, for the penetrating vehicles, that they will be in the field of view of more than one radar, and in fact many, at any time is taken for granted. Therefore, if any response is contemplated, we must identify those radars which present the greatest threat to the vehicles and initiate proper responses against them.

1.3.2 Height Finding Radars

The next electronic contact which can be expected is with a height finding radar, which generally is associated with one or more early warning radars. Because the early warning radar could usually only locate its targets in two dimensions (i.e., range and azimuth), in many cases, the defense system could be more effective if the height (elevation) of the intruder could be determined. In that case, the early warning radar could cue a *height finder* (HF) radar to the azimuth of the target of interest.

This radar is much like the early warning radar except that it has the responsibility for a much smaller volume of search (as determined by the early warning radar); it scans in elevation rather than azimuth, as does the early warning radar. Therefore, its measurement of range and elevation can be better than that of the early warning radar because its data rate is correspondingly higher. These radars are becoming obsolete with the advent of three-dimensional radars, which are designed to search in both azimuth and elevation, simultaneously or sequentially. In this case, one radar is used to perform the functions of both the EWR and HF radars.

1.3.3 Acquisition Radars

Having established that a penetrating vehicle is presenting a threat to the territories protected by the defense system, the early warning radars communicate the data as to the approximate location of the penetrators to the next line of defense, the acquisition and fire control radars. In some cases, the acquisition radar is a separate entity from the fire control radar, and in other cases, the acquisition function is included in the fire control radar.

In either case, the acquisition function is to initiate a search in the region surrounding the location identified by the early warning radar to better establish the position of the penetrating vehicle and to better identify the vehicle and its intentions. If more than one penetrating vehicle is contained within the search volume, the acquisition function must include the capability to sort between all such vehicles and determine which may be of the greatest threat to the defense system. The *field of view* (FOV) and range of detection of the acquisition radar is not as extensive as required for early warning radars; these need be no greater than required for proper guidance of the intercepting vehicles.

Although acquisition radars are used most effectively when cued by other sensors, such as the early warning radar or IR or optical sensors, they can be operated autonomously (i.e., without cueing by other sensors). Although these radars are less efficient without cueing, autonomous operation may be used as a last resort when the other sensors are not available or communication from them to the acquisition radars is disrupted. An inefficient outcome of autonomous operation of acquisition radars is more than one fire control system being deployed against the same penetrating vehicle, resulting in redundant use of ordnance by the defense system.

Although the vehicle location capability of the acquisition radar is more precise than that of the EWR, an acquisition radar is still not sufficiently accurate for precise fire control of manned or unmanned interceptors. Its primary function requires sufficient resolution and accuracy to be able to differentiate among targets when several are within the same FOV. The responsibility for more accurate location measurement and target resolution is left to the fire control radars.

1.3.4 Fire Control Radars

Fire control radars are required to locate targets of interest to sufficient accuracy to be able to guide manned or unmanned interceptors to the vehicles. Generally, the angular accuracy required is on the order of tens of milliradians; the range accuracy requirements are to within hundreds of feet. This degree of accuracy mandates a very high data rate of target location; indeed, radars which continously illuminate the target of interest are most often used; this process is referred to as *spot lighting* of the target.

In some cases, when maintaining accurate location of two or more targets simultaneously it is imperative for the fire control radar (such as a target and the intercepting missile within the same field-of-view), the radar must resort to switching the spot light between the targets. Although the data rate on each target is necessarily decreased, it must still be adequate for successful guidance of the associated interceptors. Advances in antenna designs, such as electronically steerable arrays, have enhanced the use of this capability. Most modern fire control radars currently include a multiple-target tracking capability.

1.3.5 Terminal Guidance Radars

Having successfully located its intended targets to the accuracy required with its fire control radars, the defense system launches it interceptors, manned or unmanned, toward the targets. Although guidance of the interceptors can be accomplished in many ways, this book concerns itself with two electronic forms of guidance, for example, terminal guidance with the use of an autonomous active radar on board the interceptor, or a semiactive radar which requires illumination of the targets of interest by another transmitter, usually one associated with the fire control radar (called the illuminator radar or mode).

The autonomous or *active radar* operates on the interceptor in much the same way as the fire control radar described above, using its own transmitter and receiver on the interceptor. Because of the size, weight, and cost of such a radar, this form of terminal guidance is most often found on manned interceptors that are expected to be recovered; unmanned, unrecoverable interceptors using active radars nevertheless have been and are still being used. The principal advantage associated with the use of a self-sufficient active radar on an interceptor is that the defense system can "fire and forget," which is almost imperative when two or more missiles are used to intercept separate and distinct targets from a single launching vehicle.

The principle advantage of a *semiactive radar* is that the transmitter, which is the largest, heaviest, and most costly part of a radar system is not carried aboard the interceptor; this is very desirable on those interceptors which are not expected to be recovered. The disadvantage is that the targets need to be illuminated by a cooperating transmitter, which must operate at the carrier frequency of the receiver on-board the interceptor.

The semiactive radar on-board the interceptor must also receive reference signals from the illuminating radar to be able to determine the range or range rate data contained on the signal reflected from its targets of interest. As indicated later, angular location of signals reflected from targets is determined exclusively with the antenna located at the receiver location; therefore, angular measurement is performed autonomously at the interceptor.

1.3.6 Missile Fuze

The ultimate electronic contact with the defense system would be that with the missile fuze. We should hope that one or more of the jamming techniques aboard the ECM vehicle would be effective against one or more of the previous radars, so that engagement with the defense system would not reach this stage. The missile fuze is almost always a specialized active radar system that operates on the doppler principle to determine the point of closest approach to its target.

1.3.7 Communication

Although listed and discussed last in this section, defense system communication exists during all of the electronic engagements discussed above. Communication is used for surface-to-surface weapon coordination, surface-to-air weapons guidance, air-to-surface weapon response to interrogation, and air-to-air coordination. Communication signals are continuously impinging on penetrating vehicle receiving antennas; fortunately, most of these signals are at frequencies out of the range of detection of such receivers. Nevertheless, many communication signals fall within the penetrator receiver's detection frequency range and must be processed, even though responses against these systems may not be contemplated. Electronic countermeasures against communication systems are not discussed in this book.

1.4 RADAR CONCEPTS

Before launching into discussions on the range and angle measurement techniques of the radars of interest, discussion of the concepts used by these radars to obtain that information from the signals reflected from targets would be beneficial.

1.4.1 Range Measurement

All radars base their measurement of range on the same principle: the measurement of the time elapsed from transmission of the signal to detection of the reflected signal at the radar. As shown later, because the radar signal travels at the speed of light, the time elapsed is measured in microseconds (μs, one-millionth of a second); the travel time from the radar to a target at 100 km is 665 μs, for example. This principle of range measurement is used by all types of radars, including conventional pulse, pulsed doppler, and CW doppler. The only difference among radars in range measurement is the characteristic of the radiated signal which is used as a reference in the time measurement. These differences are discussed in detail in later chapters.

1.4.2 Angle Measurement

All radars also operate on a single principle in their angle measurement function: the angle of arrival of the detected signal is defined as the pointing angle of the radar receiving antenna's main beam when detection of the reflected signal is made. This principle is based on the premise that if the reflected signal is strong enough to be detected by the radar or operator, it must have arrived via the main beam of the radar antenna, the argument being that if the reflected signal enters ouside the main beam, its power level is less than the detection level of the receiver system. This argument is supported by the fact that there can be a difference of many orders of magnitude between a signal received in the main beam and the same signal received in the antenna sidelobes. Assuming, for example, that the sidelobe level (one-way) of the radar antenna is 40 dBi, the power level of signal reflected off a target would be 80 dB less in the radar antenna sidelobes than in the main beam because of the two-way antenna loss of gain of the radar signal.

As a result, if a target detection is made by the radar circuits or the operator, we conclude that it is from a target in the main beam of the radar antenna and located in the angular direction in which the antenna is pointing. This fact must be emphasized because several of the ECM techniques discussed in later chapters take advantage of this criterion for measuring the angle of arrival of radar-detected signals.

Continuously scanning radars (i.e., search or surveillance radars, or tracking radars in their acquistion mode) operate on this same principle. The usual pointing angle accuracy of a continuously scanning radar is on the order of magnitude of a fraction of the half-power beamwidth of the radar antenna pattern. Because of the high volume of search required in these radars, the data rate is generally too low to make any more accurate measurement of the target angular position. Furthermore, the responsibility for better accuracy is left to the fire control radars, which are designed to provide a higher data rate because they are required to provide target position accuracies adequate to launch manned or unmanned interceptors towards the targets.

As shown later, the tracking radar initiates its function by pointing its antenna beam at the target of interest and making determinations as to where within the antenna main beam the reflecting vehicle is located. Whereas the search radars need only to determine where the main beam is pointing when detection of the reflected signal is made, the tracking radars are required to determine where within the main beam this reflection is detected. These more accurate angle measurements require fairly sophisticated techniques which have been designed into the tracking radar systems.

1.4.3 Range Rate Measurement

Pulsed radars measure the relative range rate (i.e., the range rate of the *line-of-sight* (LOS) distance to the target), by taking the time derivative of the measurement of several range positions. Doppler radars provide the capability of measuring the target's relative range rate by measuring the frequency shift of the carrier frequency of the reflected signal when compared to the carrier frequency of the radar transmitted signal. Although this shift is a phase shift on each reflected signal, a finite number of returns must be detected to be able to extract the shift in carrier frequency; this shift is referred to as the *doppler frequency shift.*

1.4.4 Angular Rate Measurement

Angular rate measurement can only be accomplished by successive measurements of the angle to the reflecting vehicle and subsequent differentiation of the data.

1.5 ELECTRONIC COUNTERMEASURES RESPONSE

Now that we have established the potential electronic order of battle for vehicles penetrating enemy territories, it is beneficial for us to identify typical responses which could be brought to bear against these radar defenses; this process will assure mission success and a high probability of survival. *Mission success* is defined as the probability that the mission on which the penetrating vehicle was sent will be successfully completed. In the case of a manned or recoverable unmanned vehicle, this includes the probability that the vehicle returns to its intended terminal point.

To simplify this discussion, assume that the penetrating vehicle is an aircraft flying at medium altitudes into enemy territories on a mission which will result in engagements with all of the radars discussed previously. Furthermore, assume that the aircraft is armed with an ideal receiver system (one which contains an all-band receiver with infinite sensitivity and the ability to detect and identify all of the signals impinging on the aircraft). Obviously, this is an unrealistic case, but it allows us to assume that all contacts from radars which are a threat to the aircraft are detected by the system. The main reason for this discussion is to identify possible ECM responses which could be brought to bear against the various situations that may arise during the penetration.

Research is currently underway to provide a capability as near as possible to that achieved with an ideal receiver, limited only by one or more of the three major constraints—cost, weight and volume. New technologies such as VHSIC and electronically steerable antennas are providing impetus to such receiver developments. In addition, assume that the aircraft is outfitted with an ideal ECM system (one which can respond to any and all of the radars discussed previously). Again, this type of system is not currently practical, although funds are being provided to develop a system that approximates the capabilities of the ideal system.

1.5.1 Early Warning Radar

As indicated in the previous section, the first radar contact which can be expected by a penetrating vehicle is with the EWR; unfortunately for the vehicle, this contact is expected to be with not one but many such radars, and most probably with both airborne and surface radars. An ECM designer must decide when a radiated response should be initiated, even though the designer does not know in advance whether the penetrating vehicle will be within the detection range of any of the radars.

As shown in Appendix B, because of the sensitivity of ECM receivers and the one-way path of the radar signal, the aircraft receiver will probably detect the radar signal before the radar detects the reflection from the aircraft. Therefore, a premature radiated response from the aircraft is likely to provide the early warning radar with detection capability at a range greater than will normally occur.

The desired response in this case, even if premature radiation results, is to generate signals that enter the radar's receiving circuits and prevent it from identifying the location parameters of the penetrating aircraft. As described in later chapters, this is accomplished by generating signals that conceal the true reflected signal in one or more of its coordinates, for example, range, range rate, azimuth angle, and height (in the case of a height finding or three-dimensional radar).

Concealment of a target implies an ECM signal that enters the radar receiver at virtually the same time (within nanoseconds) as the true signal return. If such a "cover" signal is not possible or not practical, the location of the true target can be concealed by radiating signals that result in a multitude of false targets at many ranges and angles, so that the probability of finding the true target position is seriously degraded.

Advanced EWRs are being designed with a capability of *looking down*, the capability of detecting reflected signals, even when immersed in very high ground reflections. In this case, the radars are forced to perform an extensive amount of signal processing to remove the naturally generated interference. This processing results in greater vulnerability of the radar to intentional interference, albeit with sophisticated techniques. These techniques are discussed at length in later chapters.

1.5.2 Acquisition Radar

Although discrete acquisition radars are implied in this section, the arguments presented apply equally well to the acquisition function of a fire control radar. As previously discussed, after detection of the penetrating vehicle, assignment is made to an acquisition radar for more accurate measurement and analysis of the location of the penetrator. Because these radars are designed with limited detection range and FOV capabilities, with a typical ECM receiver the interception of only a limited number of such radars is expected during the mission of the penetrator. However, detection of this radar generally indicates that the penetrator is within the kill distance of interceptors associated with the radar. Unless the ECM has the capability of operating against both the EWR and the acquisition radar without any degradation, it is prudent to initiate a response against the latter, even if it means a compromise in performance against the former.

Because the acquisition radar is much like the EWR (except that its detection range and FOV are only about one-third that of the EWR), the ECM response to this radar is similar to that used against the EWR. The modulation parameters used on the ECM-radiated signal against either are dictated by the peculiar characteristics of each type of radar. For example, the antenna scan rates, pulsewidths, and pulse repetition frequencies are necessarily different for the acquisition radar than for the EWR. Either cover-up signals or multiple false targets (as discussed for the EWR) are appropriate in degrading the performance of the acquisition radar.

1.5.3 Fire Control Radar

A *fire control radar* is readily identifiable by the receiver on board the penetrating aircraft because of the high data rate requirement of the radar. As a result, the radar is forced to illuminate the penetrating aircraft for a predetermined length of time, which is generally much longer than required for the early warning or acquisition radars. Because the function of this radar is to determine very accurately the position of the penetrating aircraft in order to guide an interceptor toward it, this type of radar generally receives a very high priority for ECM responses, and, unless resources are available to respond to all detected radars without loss of effectiveness against the fire control radar, ECM resources are primarily applied to respond to this radar.

Fire control radars are generally referred to as *tracking radars* because the data rate requirement designed into the radar is sufficient to allow not only very accurate location of the radar target, but also accurate prediction of the target's next position, even before the arrival of data contained in subsequent reflections from the target.

Because these radars use sophisticated techniques of radar operation to measure the location parameters of their targets, equally sophisticated ECM responses are required to confuse or deceive the radar tracking circuits. These modulations result in confusion in the radar as to the range, range rate, azimuth and elevation angle of the target position. These techniques are discussed at length in later chapters.

1.5.4 Terminal Guidance Radar

Radars used for guidance of manned or unmanned interceptors in the terminal phase of the interception are designed to operate either actively or semiactively. An active guidance radar is one which can operate autonomously, using an onboard transmitter and receiver to provide the required detection and processing. A semiactive radar is one which requires illumination of the target by another not collocated but cooperating radar; aboard the vehicle is a receiver tuned to the reflection from the target vehicle; processing of the received signal may or may not be performed aboard the vehicle.

1.5.4.1 Active Radar

If the terminal guidance on the interceptor is with the use of an active radar aboard the interceptor, that radar operates with the same process as did the fire control radar discussed earlier; that is, the terminal guidance radar provides an acquisition function with an assigned FOV that includes the region of the penetrator and subsequent lock-on and tracking of its target. Except for the motion of the missile, the signals intercepted by the ECM receiver cannot be used reliably to identify them as being from an interceptor rather than from a ground-based fire control radar. In any case, the same type of response is required against this radar as for the fire control radar.

1.5.4.2 Semiactive Radar

In the case of a semiactive guidance radar system, no signals are transmitted from the interceptor, so the ECM receiver cannot determine the location of an interceptor that is attacking the aircraft. Other sensors aboard the aircraft can be used (and usually are) to provide that information. This is one of the functions of a *tail warning radar* being installed on some of our penetrating aircraft.

Because the receiver of a semiactive radar operates in the same manner as that of a fire control radar to determine the interceptor's range, range rate, and angle relative to the penetrating aircraft, the same type of response used against the fire control radar is equally applicable against the semiactive radar. This includes confusion or deception in the radar's measurement of range, range rate, and angle.

1.5.5 Fuze

Fuze radars are unique in that they generally do not require a search function. The fixed and instantaneous FOV of the fuze radar antenna is designed so that if its intended target falls within its FOV, detonation of the warhead is indicated. Deception or confusion of this detection process is unique. Discussion of the ECM radiations required is rather sophisticated and reserved for the pertinent sections of the book.

1.6 SUMMARY

Ever since the introduction of the radar in World War II as an instrument of warfare, a continuing battle has raged between radar designers and countermeasure systems designers, the former to preserve the capabilities of the radar and the latter to deny it those capabilities. This conflict does not necessarily cease during periods of peace; it seems to intensify in preparation for any ensuing engagement.

Because of the sophistication of advanced radars, more emphasis has been placed on developing equally sophisticated techniques of ECM. Although chaff and other reflectors have been used to cause interference in victim radars, emphasis has been placed on the use of systems that deliberately radiate signals with appropriate modulations to interfere with the proper operation of the radars. An important advantage of these so-called active ECM systems is that there is no physical expenditure of resources, as with chaff and other expendable reflectors. Active ECM systems can be activated without fear of exhaustion of resources.

Active ECM systems allow extreme flexibility in the generation of interfering signals. These systems can be designed to replicate the true reflected signal to a very high degree with modulations in amplitude, phase, or frequency, which serve to confuse or deceive the victim radar. More advanced radars employing intrapulse modulations have presented formidable challenges to ECM designers; these are being met with advances in microwave signal storage and modulation techniques.

We can argue that, for every ECM technique, there is an antijamming technique; so, too, for every antijamming technique, a new ECM technique can be developed. An important aspect of antijamming or ECM design is the ease with which one or the other can be countered. Unfortunately, in too many cases, the element of surprise is an important factor in the effectiveness of either kind of technique. This is one of the reasons that advanced radar and ECM system designs have been guarded with strict security measures.

Because of the many varieties of radar types that can be expected by any vehicle penetrating enemy territory, an ECM system protecting the vehicle must be capable of countering as many of these radars as space, weight, and cost of the system allow. In addition, the ECM systems ought to be designed to attack as many of the various modes of any one radar system as the vehicle constraints allow.

Figure 1.2 depicts a typical ECM system architecture used in a modern day system. Such a system provides an active ECM capability against the various modes of different radars within the frequency range of its microwave components. Although radars of interest to the penetrating vehicle range over a frequency bandwidth of several octaves, the bandwidth for any one system is usually restricted to one octave because microwave components suffer from ambiguities and spurious signals when subjected to signals wider than one octave. Furthermore, because of the multiplicity of radars in the environment, simultaneous operation against too many can cause saturation of the microwave amplifiers and dilution of the effective radiated power of the system. Multi-octave band ECM systems are configured with parallel single octave subsystems, using common assemblies where possible.

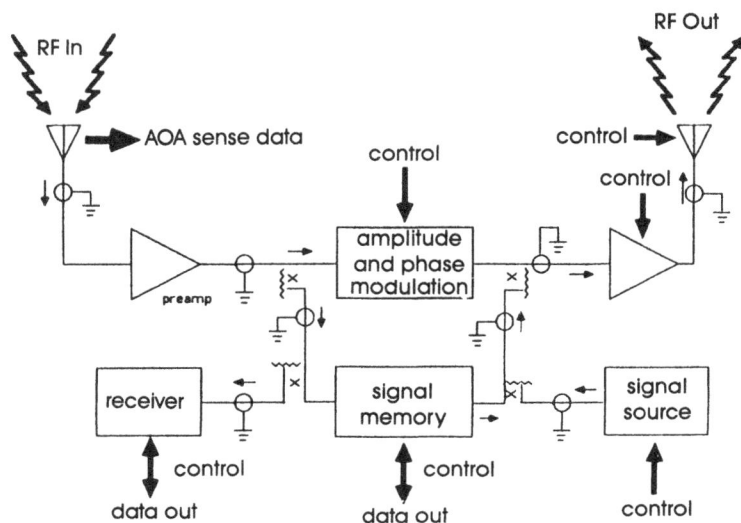

Figure 1.2 Typical ECM system architecture.

As shown in the figure, a receiver is included to detect and analyze all signals intercepted by the receiving antenna to determine which signals are associated with a radar that represents a threat to the mission or survival of the penetrating vehicle. This includes the measurement of the *angle of arrival* (AOA) of the signal so that, if necessary, the transmitted signal can be directed toward the victim radar.

By proper control of the *signal memory* and *signal source,* noise or deceptive jamming can be applied on a pulse-to-pulse basis if necessary against one or more of the radars in the environment. By appropriate control of the amplitude and phase modulator, deceptive modulations can be applied to the intercepted signal as it passes through the amplifier section of the system. Although the output amplifier of the system can be a pulse type of transmitter, most often it is a CW device that provides high duty cycle transmissions against multiple-pulse or CW radars.

Chapter 2
ECM Principles

2.1 INTRODUCTION

A radar system determines the location parameters of a vehicle by comparing the characteristics of the signal reflected from that vehicle to the characteristics of the transmitted radar signal. These characteristics are represented by the modulations contained in the transmitted radar signal and the reflected signal. As we will show in later chapters, these modulations can be in time, frequency, phase, and amplitude. The modulations contained in the reflected signal result from the unique position and motion of the reflecting vehicle relative to the radar position. Therefore, a signal deliberately transmitted from the vehicle, which is an exact replica of that impinging on the vehicle, will enter the radar receiver with the same modulation characteristics as the reflected signal. If this is so, the radar will detect the same position modulations in the transmitted signal as in the reflected signal, producing the same accurate position information at the radar as would be in the reflected signal. This is the principle used in target augmentation systems, which are designed to provide a stronger signal to the radar when located aboard a vehicle that has a very low cross-sectional area.

Therefore, an ECM transmitter aboard the vehicle generating a signal at a much greater power level than that of the reflected signals as they enter the radar receiver may not be sufficient. In most cases, the transmitter is also required to superimpose modulations on the transmitted signal that produce wrong location parameter measurements in the radar receiver. We also emphasize that it is not enough that proper deceptive modulations are superimposed on the transmitted signal; the power level of the signal must also be large enough to overcome the true modulations which occur in the reflected signal.

As we indicate later in this chapter, the amount of power reflected from the vehicle is directly proportional to the radar cross-sectional area of the vehicle as viewed on the radar line of sight (LOS) to the target. This cross-sectional area is not necessarily the physical cross section of the vehicle, but is related to it. Although much effort is being expended to reduce this reflective area (called stealth or radar cross section reduction programs), we do not expect that it will be reduced to a level such that zero power is reflected from the vehicle. Nevertheless, the reduction being achieved can seriously degrade the performance of existing radars and force the enemy to improve its radars to be able to detect the lower power levels caused by the reduced cross-sectional area.

Therefore, we conclude that a reflected signal from the protected vehicle, however small, will be available at the antenna terminals of the radar; whether the power level of the reflected signal is great enough to be detectable depends on the performance characteristics of the radar. Because no ECM technique has yet been devised to prevent this signal from entering the radar receiver, we must also conclude that an ECM system must compete with the signal in power level so as to deny the radar the true vehicle location parameters as contained in the modulation of the reflected signal.

However, as discussed above, that the ECM signal conceals the characteristics of the reflected signal is not sufficient; modulations must also be imposed on the ECM signal that conceal the position of the protected vehicle. The next and following chapters discuss at length the nature and the types of modulations required on the ECM transmitted signal to confuse or deceive the radar as to the true position of the protected vehicle.

The objective of this chapter is to develop formulas that can be used to determine the power level of the signal reflected from penetrating vehicles when illuminated by hostile radars as well as formulas to define the power level of signals radiated from ECM systems designed to protect these vehicles. From these formulas, others are generated to define the ratio of the jamming signal to the reflected signal as a function of the performance characteristics of the radar and ECM systems, as well as the position of the reflecting vehicle relative to the radar of interest. This ratio is known as the *jamming-to-signal ratio* (*J/S*).

These *J/S* determinations are developed independently of the modulations used or proposed in subsequent chapters. In subsequent discussions of modulation techniques provided in ECM systems, *J/S* values are suggested as being required for effective operation with each jamming technique. With the background of this chapter, we can develop the performance parameters required in an ECM system designed to provide that capability. This approach is taken to eliminate the requirement to determine the ECM performance characteristics each time a new technique is described. Even so, the required *J/S* values are provided in each case (as well as ECM performance characteristics) without burdening the reader with the required calculations for each case. The interested reader can perform the calculations independently using the formulas developed in this chapter.

2.2 REFLECTED SIGNAL POWER LEVEL

As shown in Figure 2.1 the signal, in this case a pulse, is radiated by the radar

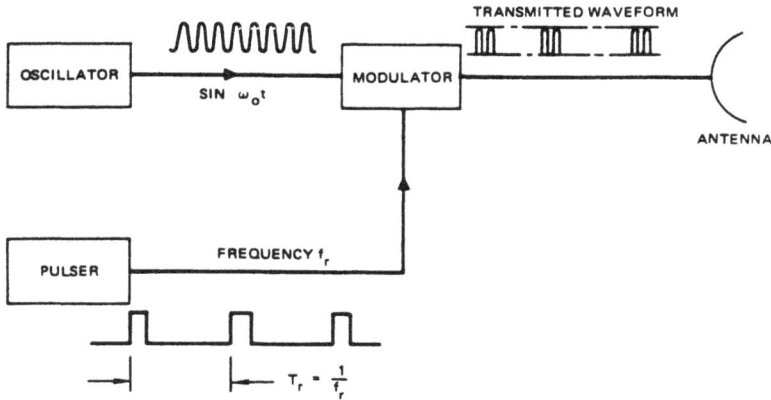

Figure 2.1 Radar transmission of pulse waveform (from [1]).

and directed toward the aircraft. The radiation from the radar impinges on the vehicle and is reflected, with some of the reflection returning over the same path toward the radar. The power level of this returned signal at the radar is determined in this section as the S value of J/S which ultimately needs to be determined.

Figure 2.2 shows how a single antenna serves as both a transmitting and receiving device.

As shown in Figure 2.3, assume the radiation from a radar transmitter is from a point source whereby all power transmitted, P_t, radiates in all directions. In this case, the power density at any point in range is equal to the total power distributed over a sphere whose surface area is equal to $4\pi r^2$, where r is equal to the distance between the point source and the sphere surface. The power density is then

$$S = \frac{P_t}{4\pi r^2} \cdots \tag{2.1}$$

where the dots in the formula indicate that the equation is incomplete and that further development of the equation follows.

If the radar transmits via an antenna which has directivity and generates maximum power in the direction of the aircraft, rather than radiating equally in all directions (isotropic antenna) as in the previous case, the power density in the direction of the aircraft must be modified by the gain of the antenna in that direction, as shown in Figure 2.4. Obviously when power is concentrated in a particular direction because of the directivity of the antenna, the integrated gain

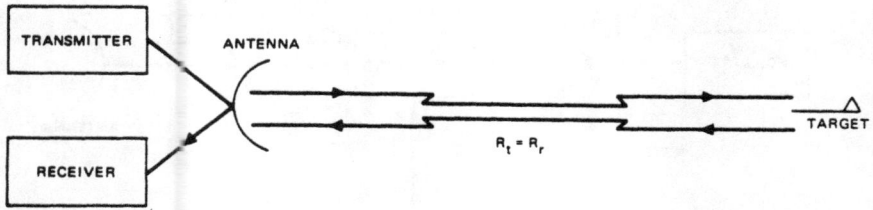

Figure 2.2 Single transmitting and receiving antenna (from [1]).

$$S = \frac{P_t}{4 <\pi> r^2} \cdots$$

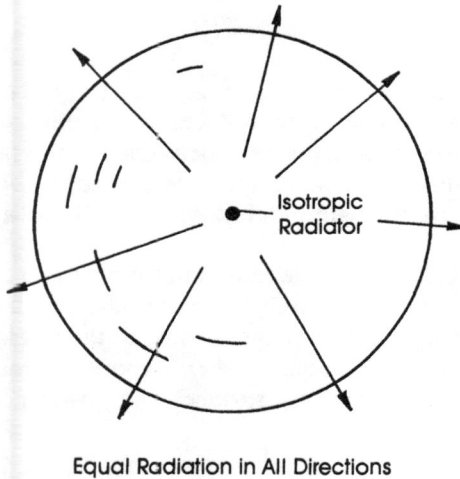

Equal Radiation in All Directions

Figure 2.3 Radiation from isotropic radiator.

in all other directions must be lower than it is for an isotropic antenna. Assuming that the gain in the direction of the aircraft is G_t, the S equation is modified to

$$S = \frac{P_t G_t}{4\pi r^2} \cdots \tag{2.2}$$

Eventually, the expanding radiation (increased r) reaches the target. As the signal impinges on the aircraft, its power density is as shown in (2.2). The amount of power reflected at the vehicle is then determined by multiplying the power density

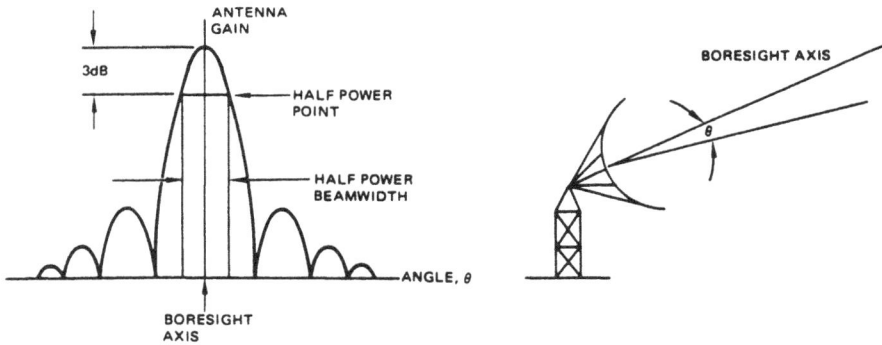

Figure 2.4 Typical antenna gain curves (from [1]).

by the effective *radar cross section* (RCS) of the aircraft. As shown in Figure 2.5, the RCS of the typical vehicle is very erratic, varying randomly at the various aspect angles to the vehicle. Often the RCS value used in calculations is the average value of those about the angle of interest, with little loss in generality, because radar receivers tend to smooth out the random variations in reflective area. Table 2.1 provides typical average values for the RCS of vehicles of interest. Giving the RCS the notation σ, the S equation now becomes

$$S = \frac{P_t G_t}{4\pi r^2} \sigma \ldots \tag{2.3}$$

At this point, the equation defines the amount of power reflected off the vehicle, and, as shown in Figure 2.5, the power level is dependent on the radar aspect angle to the vehicle. Figure 2.6 shows the smoothed characteristics of typical RCS of an aircraft type of target as a function of aspect angle. Assume that this amount of power is radiated equally in all directions, as was done with the radar power transmitter in Figure 2.3 so that the power density of the reflected signal is dependent on the surface area of the radiation sphere about the reflector and which is dependent on the range from the reflecting vehicle and the radar. The S equation (Figure 2.7) now becomes

$$S = \frac{P_t G_t \sigma}{4\pi r^2} \cdot \frac{1}{4\pi r^2} \ldots \tag{2.4}$$

As this radiation sphere impinges on the radar antenna, the amount of power

Table 2.1

Mean Radar Cross Section of Typical Targets at 1.3–10 GHz (Courtesy of Technology Service Corp.)

Aircraft (Nose and Tail Aspect)	
Small General Aviation	0.6–3 m^2
Small Fighters	1.5–4 m^2
Medium Fighters (F-4, *et cetera*)	4.0–10 m^2
Small Commercial (DC-9)	10–20 m^2
Medium Commercial (707, DC-8)	20–40 m^2
Large Commercial (DC-10, 747)	40–100 m^2
Ships (5–10 GHz; Approximate Frequency Dependence = $f^{1/2}$)	
Sailboats — Small	0.5–5 plus mast
Military Power Boats	20–500
Frigates (1–2 ktons)	0.5–1.0 × 10^4
Destroyers (3–5 ktons)	3.0–6.0 × 10^4
Cruisers (7–20 ktons)	10–40 × 10^4
Carriers (20–40 ktons)	30–100 × 10^4
Tanks	20–200 m^2
Personnel	0.3–1.2 m^2
Birds	
Sparrow and Starling	.0003–.001 m^2
Pigeon (25–40 knots)	.001–.01 m^2
Mallard (25–40 knots)	.01 m^2

captured by the antenna depends on the effective cross-sectional area of the antenna, which is given as A_t. The total reflected power at the radar antenna terminals (Figure 2.8) is therefore

$$S = \frac{P_t G_t \sigma}{(4\pi)^2 r^4} \cdot A_t \tag{2.5}$$

The effective cross-sectional or *capture area* of an antenna is given as

$$A_t = \frac{G_t \lambda^2}{4\pi}$$

where λ is equal to the wavelength of the radar carrier.

Replacing A_t in the S equation thus results in what is called the *radar equation*, which is used to calculate the power received at the radar antenna terminals from a reflecting target of size, σ, at a range of r due to illumination of the target by a radar transmitter, the power level of which is P_t, which is transmitted via a directive antenna with a gain of G_t, in the direction of the target. The radar equation, then, is

$$S = \frac{P_t}{4 <\pi> r^2} \times G_t \times <\sigma> \dots$$

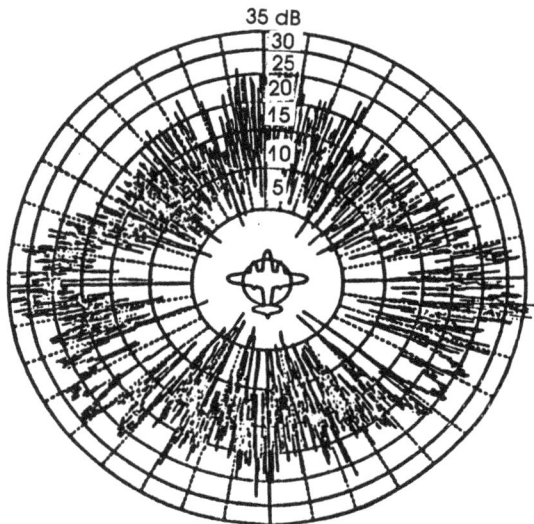

Figure 2.5 Aircraft radar cross-section variation.

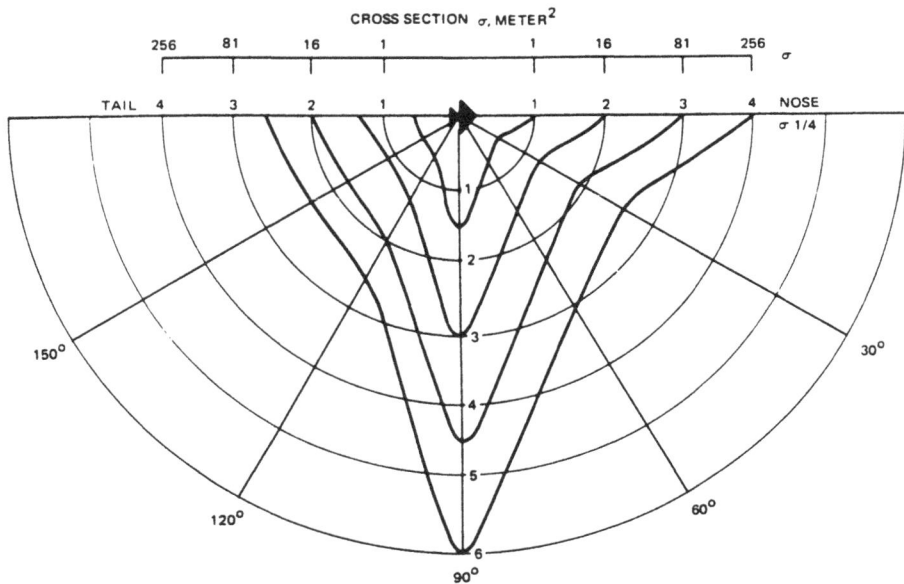

Figure 2.6 Smoothed aircraft cross-section (from [1]).

$$S = \frac{P_t}{4<\pi>r^2} \times G_t \times <\sigma> \frac{1}{4<\pi>r^2} \cdots$$

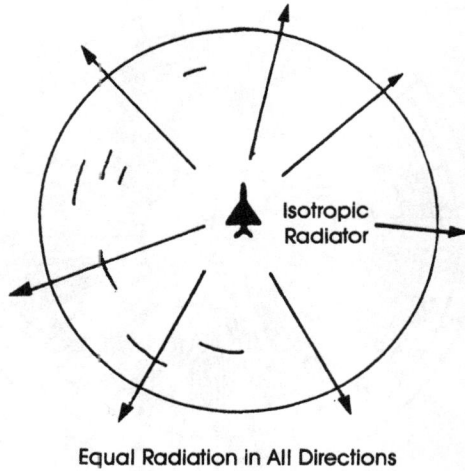

Equal Radiation in All Directions

Figure 2.7 Reflection from vehicle.

$$S = \frac{P_t}{4\pi\,r^2} \times G_t \times \sigma \times \frac{1}{4\pi\,r^2} \times A_t$$

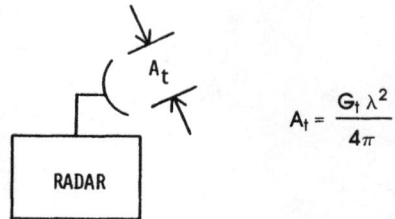

RADAR

$$A_t = \frac{G_t\,\lambda^2}{4\pi}$$

Figure 2.8 Antenna aperture *versus* gain.

$$S = \frac{P_t G_t^2\,\sigma\lambda^2}{(4\pi)^3\,r^4} \tag{2.6}$$

An important facet of this equation is that the radar received power level is inversely proportional to the 4th power of the range to the reflector; this is because of the two-way path of the radar signal. This is especially important because, as we will show later, the ECM signal is on a one-way path so that the ECM power level as received by the radar is inversely proportional to only the second power of the radar range to the target. As an example of this equation, let us assume the following parameters:

P_t = 1 MW
G_t = 35 dBi
σ = 10 m^2
λ = 10 cm (3 gHz)
r = 10 km

Therefore, $S = -43$ dBm. (Appendix A presents a review of the use of the decibel (dB) for the interested student.)

Assuming S_1 equals the received power level when the reflecting target is at r_1, and S_2 is the received power level when the target is at r_2:

$$S_1 = \frac{P_t G_t^2 \sigma \lambda^2}{(4\pi)^3 (r_1)^4}$$
$$S_2 = \frac{P_t G_t^2 \sigma \lambda^2}{(4\pi)^3 (r_2)^4}$$
$$S_2/S_1 = [r_1/r_2]^4 \tag{2.7}$$

Equation (2.7) shows that the ratio of received power levels varies as the fourth power of the ratio of the two target ranges. In decibels, this is shown to be

$$S_2 = S_1 + 40(\log[r_1/r_2]) \tag{2.8}$$

Equation (2.8) is an equation for a straight line on a semilogarithmic graph where S is the linear y-coordinate and r is the logarithmic x-coordinate as shown in Figure 2.9.

If r_1 is 10 km as above and r_2 is 100 km, the received power level from a target at 100 km can be calculated as

$$S_2 = S_1 + 40[\log (10/100)] = -43 - 40 = -83 \text{ dBm}$$

The above data give two points on the straight line on the semilogarithmic graph and allows us to draw the line as in Figure 2.9. Indeed, it was possible to draw this line given the one value at r and the slope of the line, as shown in (2.8), to be -40 for each tenfold change in range.

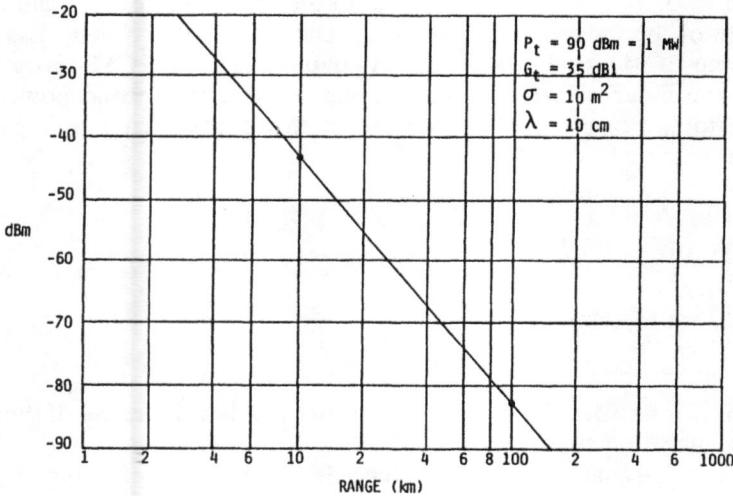

Figure 2.9 Return signal power *versus* radar range.

The line on the graph in Figure 2.9 was drawn with two easily determined points. This line makes possible determination of the received power level for this particular radar and a target at any range. Indeed, all solutions for any value of P_t, σ, λ, and G will result in lines parallel to the one shown, except that they will be displaced in the y-coordinate by an amount dependent on the ratio of the new values relative to that used in this example. For example, for a radar with a tenfold increase in power level compared to the example (or a tenfold increase in σ), the solution will result in a line parallel to the one given in the example, but increased in the y-coordinate by 10 dB.

2.3 POWER RECEIVED AT THE RADAR FROM A CONSTANT POWER TRANSMITTER

Figure 2.10 depicts the case of an airborne transmitter radiating toward a radar. We can see immediately that this signal travels only one way from the transmitter to the receiver in contrast to the radar signal, which travels two paths, one from the radar to the target and the second from the target to the radar. Using the same reasoning as for the radar signal, we can show that the signal at the radar antenna terminals from a transmitter aboard the target is as follows:

$$J = \frac{P_j G_j G_t \lambda^2}{(4\pi)^2 r^2} \tag{2.9}$$

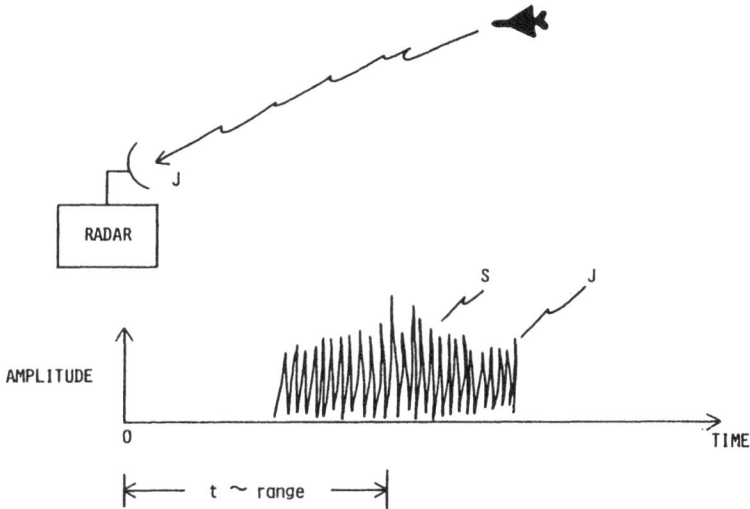

Figure 2.10 Noise jamming of pulse radar.

where

P_j = target transmitter power
G_j = target transmitter antenna gain

In this case, the equation shows that the received power is inversely proportional to the square of the range from the radar to the target. For,

P_j = 100 W
G_j = 5 dBi
G_t = 35 dBi
r = 10 km

Therefore, $J = -32$ dBm. We can also show that

$$J_2 = J_1 + 20 \, (\log[r_1/r_2])$$

For convenience, we use r_2 equal to 100 km, which yields

$$J_2 = J_1 - 20 = -52 \text{ dB}$$

As for the radar equation, Figure 2.11 shows the line which results from the jamming equation with the parameters given. Note that the slope of the line is -20 dBm per decade, whereas it is -40 dBm per decade for the radar equation.

2.4 JAMMING-TO-SIGNAL RATIO (J/S)

Thus far, we have developed the equations for the power level received by a radar from a reflecting target as well as the power level received by the same radar from a jamming system located at the target. We shall now develop the equation which is of most importance to this book, that which derives the ratio of the jamming signal to the target echo signal.

The ratio of (2.9), J, to (2.6), S, is

$$J/S = \frac{P_j G_j}{P_t G_t} \cdot \frac{4\pi r^2}{\sigma}$$

(2.10)

Using the parameters given in the above examples for the radar and the jamming system, the calculation of this equation for r equal to 100 km results in

$$J/S = 31 \text{ dB}$$

This states that the received power level from a jammer, with the specified parameters, from a target at 100 km is 31 dB greater in power than the power level of the reflected signal received by the radar.

We can easily see this same result if we draw the radar line shown in Figure 2.9 and the jamming line on the same graph shown in Figure 2.11. This is done in Figure 2.12. On this graph, we can see that the J/S relationship at 100 km is 31 dB, as calculated. With this graph, it is possible to determine the J/S values at any range for the given set of parameters. As we can see, J/S decreases as the range to the target decreases; this is because, with decreasing range, the target signal increases at a faster rate than does the jamming signal.

2.5 BURN-THROUGH RANGE

As we can see in Figure 2.12, because of the differences in the range relationships of the radar signal and the jamming signal (and thus the slopes of their respective lines), the two lines eventually must intersect. This occurs when the two power levels are equal ($J/S = 1$); the range at which this occurs for any one set of parameters (radar and jammer) is often referred to as the *jammer crossover range* and in some cases as the *burn-through range*. The latter nomenclature was applied because it was believed that the ability of an operator to distinguish between a

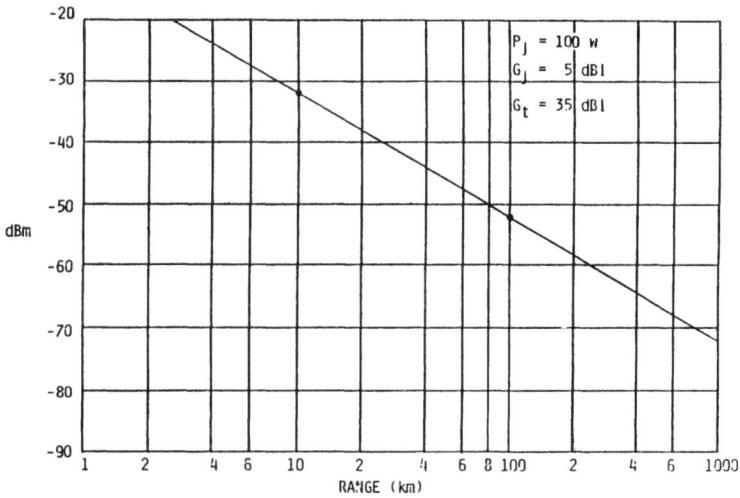

Figure 2.11 Jamming power *versus* radar range.

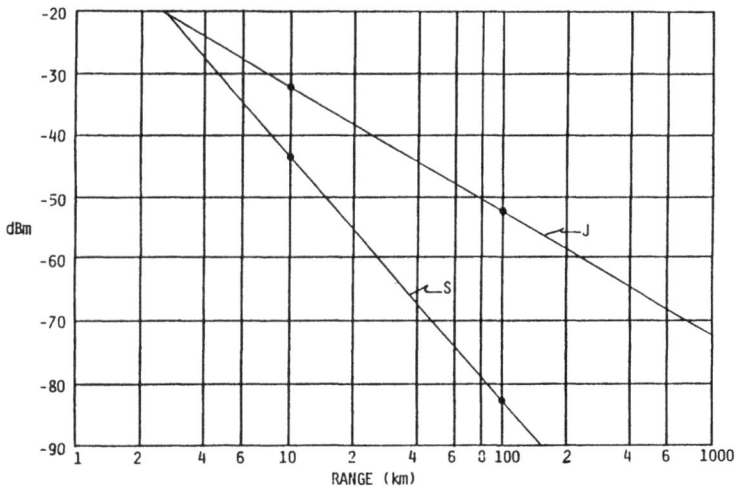

Figure 2.12 Jam-to-signal ratio *versus* radar range.

true target signal and the noise of a jammer on a video display occurred when the J/S was equal to 1. Recently, the term "burn-through" has been more generally applied; it is now also used to indicate the J/S value at which the radar (operator or radar circuits) can reliably detect the true signal over the interference. This detection can vary widely; in some cases, this can be accomplished with J/S values as high as 10 dB and, and as low as $-$ 10 dB.

The graph in Figure 2.12 allows determination of the burn-through range for any amount of specified J/S. For example, if burn-through range is required for a J/S value of 10 dB, we only need to find the range at which the distance between the two lines is 10 dB; in this case, this occurs at approximately 9 km. For the crossover ($J/S = 1$) burn-through, the range is approximately 3 km.

Using the equations, we can show that the range at which J/S equals any value can be calculated from

$$r = \left[\frac{P_t G_t}{P_j G_j} \cdot \frac{\sigma}{4\pi} \cdot \frac{J}{S} \right]^{1/2} \tag{2.11}$$

for $J/S = 1$, $r = 2.2$ km, which verifies the value read from the graph.

2.6 JAMMER POWER REQUIRED

For proper design of a jamming system, we must determine the transmitter power required for effective operation of the jammer. The required power level is usually defined as the *effective radiated power* (ERP) of the system, and is defined as the product of the transmitter power at the jammer transmitter antenna terminals and the gain of the jammer antenna in the direction of the victim radar; this is equal to $P \times G$, where P is the transmitter power at the antenna terminals and G is the directive gain of the transmitting antenna. Rearranging (2.11), we find

$$P_j G_j = P_t G_t \cdot \frac{\sigma}{4\pi r^2} \cdot \frac{J}{S} \tag{2.12}$$

As shown in (2.10) and the associated graph, J/S decreases in value as the range from the jammer to the radar decreases, which dictates that the range to be used in (2.12) is the minimum range at which the stated J/S is required. As shown in the figures, at ranges greater than this minimum range, the J/S value increases; this is also evident in (2.10). Therefore, a jammer is effective at long ranges, but decreases in effectiveness as the target range to the radar decreases.

As an example, determine the ERP required by a jamming system to provide a J/S of 31 dB at a range of 100 km against a radar with the characteristics given above. This is calculated to be 300 W, the ERP postulated for the jammer in the previous calculations, which yielded a 31 dB J/S at 100 km.

2.7 POWER DILUTION DUE TO JAMMING BANDWIDTH

Thus far, we have assumed that all jamming power was contained within the detection bandwidth of the victim radar receiver. In most cases, the jamming power is transmitted over a wider bandwidth than the detection bandwidth of the radar receiver, usually for one of two reasons. The first is the inability of receivers associated with jamming systems to measure and reproduce accurately the operating frequency of the radar signal. The other reason is the requirement for simultaneous operation against more than one radar, with radars operating at different frequencies.

In the first case, the jammer transmits over a relatively wide frequency range to compensate for the inaccuracy of the frequency measurement. In the second case, a wide frequency range is transmitted to include the frequencies of all radars of interest at the same time. Equation (2.12) must then be modified to allow for this dilution of ERP because jamming power transmitted outside the detection bandwidth in the radar receiver does not contribute to disruption of the radar detection process. As shown in the equation below, this modification is simply a multiplication of the equations above by the ratio of the jamming bandwidth to the detection bandwidth. This is shown symbolically in Figure 2.13. When the jamming bandwidth is less than the radar detection bandwidth, the bandwidth ratio is taken as equal to 1.

$$P_j G_j = P_t G_t \cdot \frac{\sigma}{4\pi r^2} \cdot \frac{J}{S} \cdot \frac{B_j}{B_r} \qquad (2.13)$$

2.8 JAMMER LOCATED AT DIFFERENT RANGE FROM TARGET

The previous equations were also based on the assumption that the jamming system was located on the vehicle being protected by the jammer. There are situations where it may be advantageous to locate the jammer at a position different from that of the protected vehicle (in range as well as in angle). In that case, (2.13) must be modified to account for this difference. The following discussion determines the modification required for differences in range only. The next section discusses the required modifications to the equations when the jammer and the vehicle being protected are at different angles relative to the radar. The modified equations are found by using the formulas for radar received power from the target separately from the radar received power from the jammer, as follows:

$$P_{rj} = \frac{P_j G_j G_t \lambda^2}{(4\pi)^2 r_j^2} = J$$

NOISE POWER DILUTION:

$$\frac{J}{S} = \frac{P_j\, G_j}{P_t\, G_t} \times \frac{4\pi\, r^2}{\sigma} \quad \cdots\cdots$$

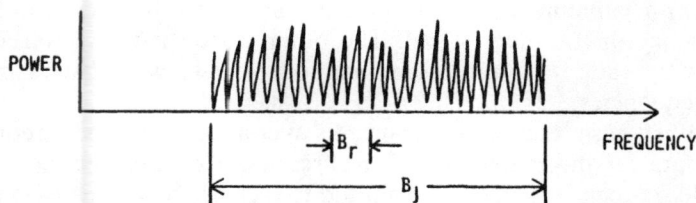

POWER | FREQUENCY

$$\frac{J}{S} = \frac{P_j\, G_j}{P_t\, G_t} \times \frac{4\pi\, r^2}{\sigma} \times \frac{B_r}{B_j}$$

$$P_j\, G_j = P_t\, G_t \times \frac{\sigma}{4\pi\, r^2} \times \frac{J}{S} \times \frac{B_j}{B_r}$$

Figure 2.13 Noise power dilution.

where r_j = range to the jamming vehicle, and

$$P_{rt} = \frac{P_t G_t^2 \sigma \lambda^2}{(4\pi)^3 r_r^{\epsilon}} = S$$

where r_t = range to the target. Thus,

$$J/S = \frac{P_j G_j}{P_t G_t} \cdot \frac{4\pi}{\sigma} \cdot r_t^2 \cdot \left(\frac{r_t}{r_{j}}\right)^2 \tag{2.14}$$

Also,

$$P_j G_j = P_t G_t \cdot \frac{\sigma}{4\pi r_t^2} \cdot \frac{J}{S} \cdot \left(\frac{r_j}{r_t}\right)^2 \tag{2.15}$$

We can see from the above equations that the modification to the previously developed equation for J/S consists of a multiplication of the J/S equation by the square of the ratio of the range to the target to the range to the jammer. In the case of the ERP equation, a multiplication by the square of the inverse ratio is required. When the ECM system is located aboard the vehicle being protected, $r_j = r_t$, and the basic jamming equation results.

2.9 JAMMER LOCATED AT DIFFERENT ANGLE FROM TARGET

We emphasize that in all of the cases cited above we assumed that the target and jammer were at the same angle to the radar, even in the noncollocated jammer. This section describes the case when the jammer and target are not located within the same beamwidth of the radar antenna.

The modified equations are found by using the formulas for radar received power from the target separately from the radar received power from the jammer, as follows:

$$J = \frac{P_j G_j G_{SL} \lambda^2}{(4\pi)^2 r_j^2}$$

where r_j = range to the jamming vehicle and G_{SL} = radar antenna sidelobe gain in the direction of the jamming vehicle, and

$$S = \frac{P_t G_t G_{ML} \sigma \lambda^2}{(4\pi)^3 r_t^4}$$

where r_t = range to the target and G_{ML} = radar antenna mainbeam gain in the direction of the target. Thus,

$$J/S = \frac{P_j G_j}{P_t G_t} \cdot \frac{4\pi}{\sigma} \cdot r_t^2 \cdot \left(\frac{r_t}{r_j}\right)^2 \cdot \frac{G_{SL}}{G_{ML}} \tag{2.16}$$

Also,

$$P_j G_j = P_t G_t \cdot \frac{\sigma}{4\pi r_t^2} \cdot \frac{J}{S} \cdot \left(\frac{r_j}{r_t}\right)^2 \cdot \frac{G_{ML}}{G_{SL}} \tag{2.17}$$

We can see from the above equations that the modification to the previously developed equation for J/S consists of a multiplication of the equation by the ratio of the radar antenna sidelobe gain to the radar antenna main beam gain. In the case of the ERP equation, a multiplication by the inverse ratio is required. When the ECM system is located at the same angle as the target being protected, $G_{SL} = G_{ML}$, and the basic jamming equation results.

2.10 CONSTANT POWER *VERSUS* CONSTANT GAIN SYSTEMS

Figure 2.14 shows the block diagram of a typical multimode all-purpose jamming system. RF signals intercepted by the receiving antenna at (a) are preamplified and, at first interception, are channeled to the receiver via the directional couplers shown. These signals are analyzed as to their importance to the survival of the host vehicle, and a decision is made as to the response which would provide the greatest protection to the vehicle. If this response is an active ECM response, a decision is made as to what type of transmitted signal and modulation are required for the desired response.

2.10.1 Repeater—Constant Gain System

Figure 2.15 indicates the portions of the system architecture which are used to provide the repeater function. Assuming that the signal to be transmitted is to be a true replica of the signal intercepted, the signal from the receiving antenna is passed through the amplitude and phase modulation unit, which provides the appropriate modulation to the signal as it passes through the circuit. The output of this circuit is passed through the output amplifier and the transmitting antenna toward the victim radar. This mode of operation is referred to as the *repeater* mode because the signal transmitted is coherently derived from the signal intercepted at the ECM receiving antenna. Deceptive modulations are superimposed on this signal to confuse or deceive the radar analysis circuits.

In this type of operation, the output of the transmitter is directly proportional to the intercepted signal; this type of system is referred to as a *constant gain* system. The output of the transmitter is not necessarily the maximum output of the transmitter, but it is dependent on the level of the intercepted signal multiplied by the gain of the amplifying system. The gain of the system to the intercepted signal must be less than the feedback attenuation (isolation) between the transmitting antenna and the receiving antenna. This requirement exists because the transmitter is radiating almost simultaneously as the intercepted signal is being received at the receiving antenna. If the gain of the system were greater than the feedback attenuation, the feedback signal would be greater than the intercepted signal; this situation could cause oscillation in the amplifying system. For this reason, the power output of the transmitter is designed to be directly proportional to the intercepted signal power level and, in most cases, is less than its maximum power output capability.

2.10.2 Transponder—Constant Power System

Figure 2.16 depicts the portions of the system architecture which are used to provide the transponder capability. The response decision may require modulation of the

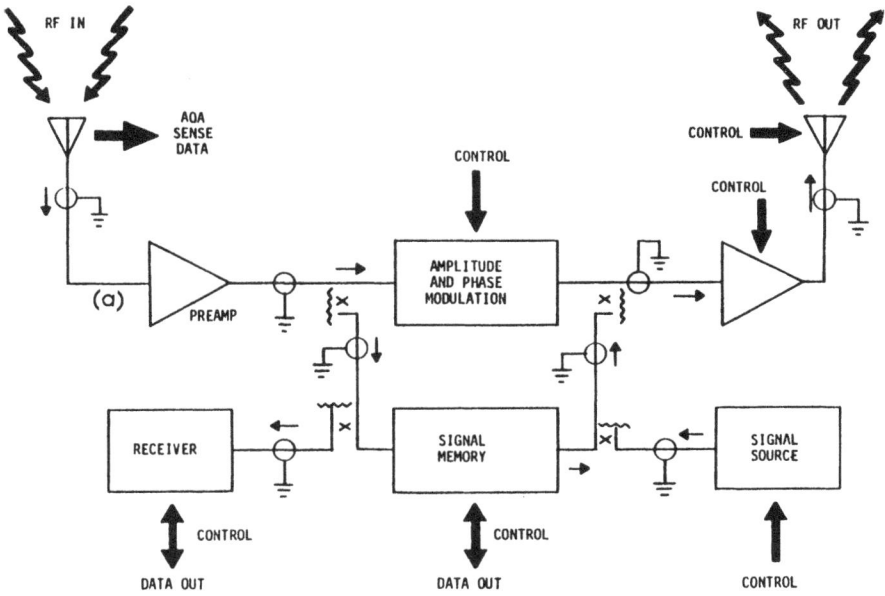

Figure 2.14 Typical ECM system architecture.

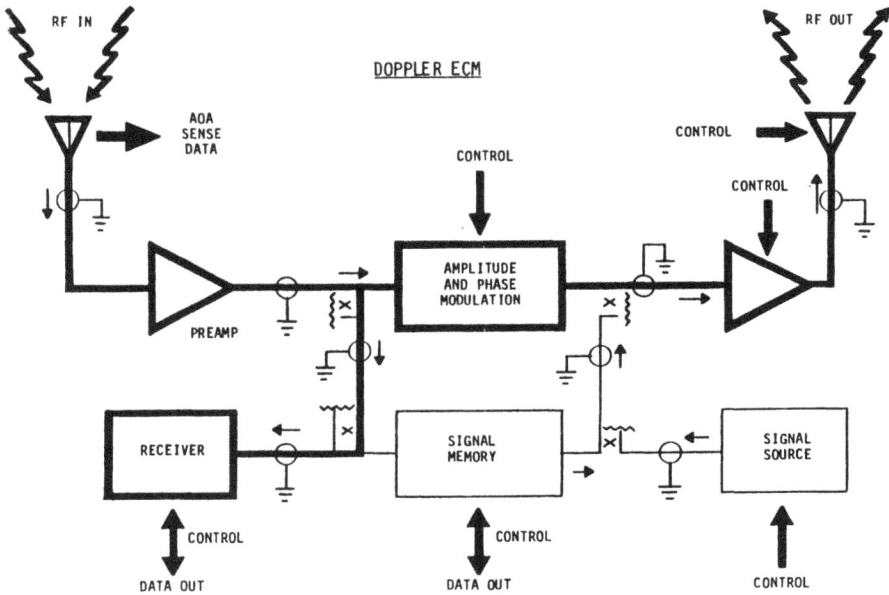

Figure 2.15 Repeater configuration.

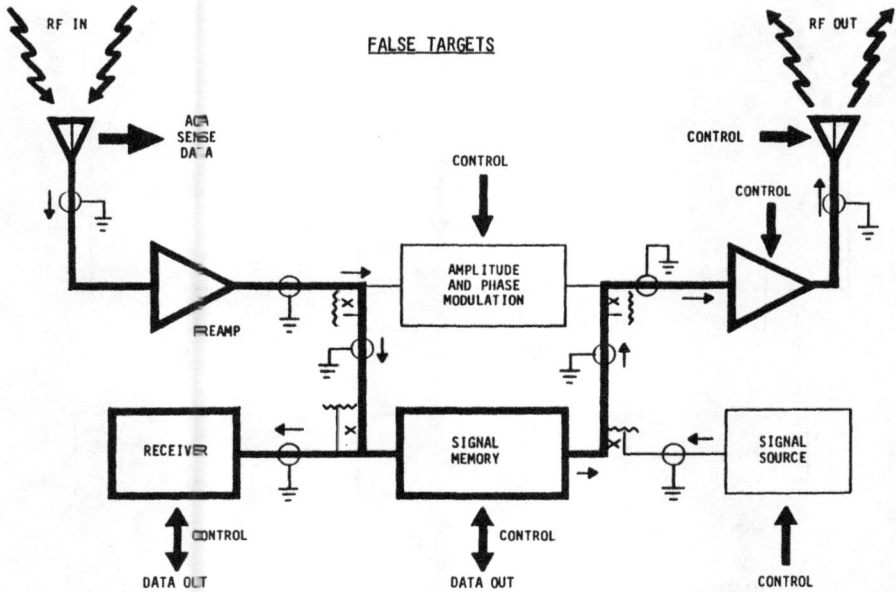

Figure 2.16 Transponder configuration.

signal in time (i.e. with a time delay between the time of signal interception and signal transmission). The effect of this modulation on radar receivers is discussed in later chapters. In this case, it is necessary to store the intercepted signal in some manner in order to make it available for transmission well after the incoming signal has terminated at the input to the system. Much effort has and is being expended to develop technology capable of storing this signal as coherently as possible in order to provide an accurate replica of the signal intercepted.

We define *coherence* as the accuracy with which the intercepted signal can be reproduced in its carrier frequency, which includes any frequency or phase modulation contained within the intercepted signal. Although, ideally, exact frequency and phase coherence is desired (the rigorous definition of coherence), reproduced frequency accuracy to within tens of hertz during the duration of the intercepted signal has been shown to be acceptable.

Usually the microwave storage system reproduces the intercepted signal at a power level that is required to produce maximum power out of the transmitter independent of the signal level of the intercepted signal. This results in a constant power level out of the system, which justifies references to this type of operation as a *constant power system*. This mode of operation is acceptable, even when the amount of isolation between the transmitting and receiving antennas is not adequate to prevent a feedback signal greater than the intercepted signal because, in normal operation, the ECM receiver is gated off during transmission of the ECM signal.

2.10.3 Noise—Constant Power System

The ECM response desired may be a noise source with appropriate modulation to confuse or deceive the victim radar. Figure 2.17 depicts the parts of the general block diagram which are used to accomplish that function. If analysis of the intercepted signal by the receiver indicates that an appropriately modulated noise source is most effective in the current situation, a signal source as near as possible to the carrier frequency of the victim radar is used, with appropriate modulation. Because the power level of this noise source is generally independent of the intercepted signal, it is adjusted to provide maximum power out of the transmitter at all times, except for required jamming modulation. This type of operation is one form of a constant power system.

2.11 CONSTANT GAIN SYSTEM CALCULATIONS

To determine the gain required in a constant gain system, let us use (2.10):

$$J/S = \frac{P_j G_j}{P_t G_t} \cdot \frac{4\pi r^2}{\sigma}$$

$$P_j = P_i G_A \tag{2.18}$$

where

$$P_i = \text{intercepted signal power level}$$

$$G_A = \text{repeater amplifier gain}$$

$$P_i = \frac{P_t G_t}{4\pi r^2} \cdot \frac{G_i \lambda^2}{4\pi} \tag{2.19}$$

where G_i is the repeater receive antenna gain.

$$J/S = \frac{P_t G_t G_i \lambda^2}{(4\pi)^2 r^2} \cdot G_A \cdot \frac{G_j}{P_t G_t} \cdot \frac{4\pi r^2}{\sigma}$$

$$J/S = G_i G_A G_j \cdot \frac{\lambda^2}{4\pi\sigma} = G \cdot \frac{\lambda^2}{4\pi\sigma}$$

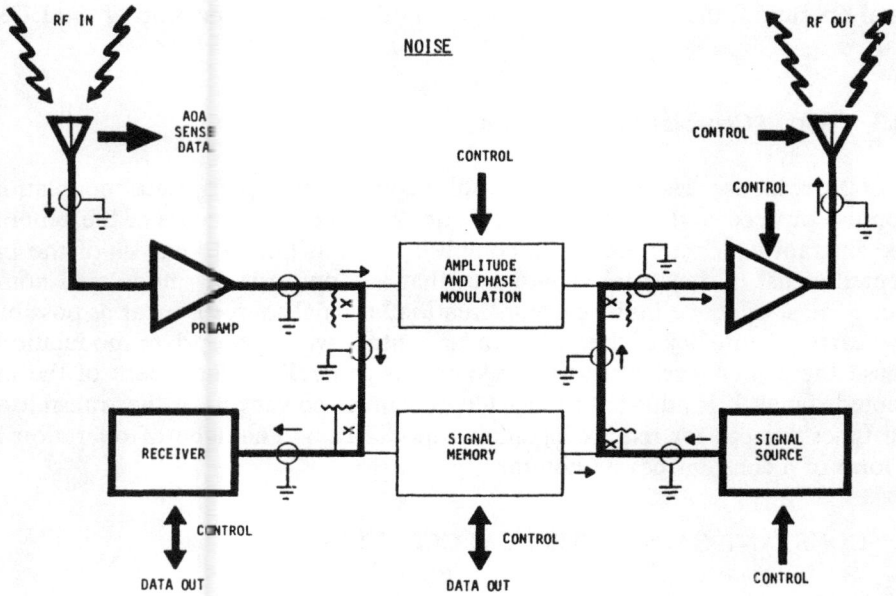

Figure 2.17 Noise configuration.

where $G = G_i G_A G_j$. Finally,

$$G = \frac{4\pi\sigma}{\lambda^2} \cdot \frac{J}{S} \tag{2.20}$$

This equation states that the gain of a constant gain system required to provide a predetermined J/S value is equal to product of the gain of the ECM system receiving antenna, the gain of the amplifying system, and the gain of the ECM transmitting antenna, and is directly proportional to the RCS of the vehicle being protected and the J/S value required.

For example, for a J/S value of 10 dB and an RCS value of 10 m^2 at a radar carrier wavelength of 10 cm, we have

$$G = \frac{4\pi\sigma}{\lambda^2} \cdot \frac{J}{S}$$

$$G = 51 \text{ dB}$$

The line which represents this type of operation on the semilog graph used for the radar and ECM signal is easily drawn by noting that at all ranges the transmitter output power level is a constant value (*J/S*) above that calculated for the echo return power. Therefore, a line parallel to the echo signal line, but displaced in the *y* direction by an amount equal to this *J/S* value, represents the line required for constant gain operation. This line is shown in Figure 2.18 for the example above, which requires a *J/S* value of 10 dB.

2.11.1 Limiting Value of a Constant Gain System

As shown in Figure 2.18, assuming the ECM transmitter was not limited in its power capabilities, the solid line 10 dB above the echo power would continue indefinitely parallel to that line. However, because transmitter systems are (necessarily) designed with limited power levels, the gain of the system eventually will drive the level of the signal to that which demands maximum power out of the transmitter. Having reached this power level, any further increase in intercepted signal will result in the same maximum power level out of the transmitter; at ranges less than that point, the system operates as a constant power system.

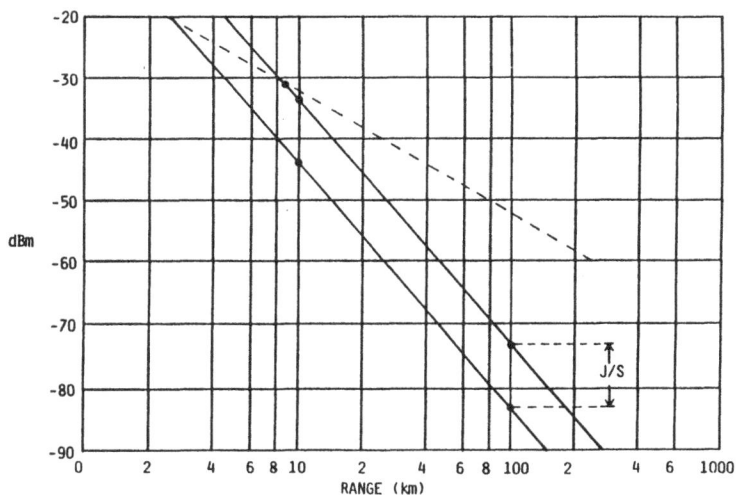

Figure 2.18 Constant gain system graph.

The dotted line in the figure represents the power received by the radar if the ECM system in the example is operating as a constant power system. As we can see, at long ranges, the power received by the radar from a constant gain system is well below the capabilities of the ECM transmitter. This condition is forced on the ECM system because of the gain limitations of a constant gain system, due to lack of antenna isolation. As seen in the figure, at some range and thus at some input signal level, the transmitter power level out of the constant gain system is exactly equal to what the power level out of the system would be if it were operating as a constant power system. In Figure 2.18, this we see as a range of about 9 km. From that point in, the system operates as a constant power system and the line representing a constant gain system proceeds along the constant power line. In either case, therefore, the crossover ($J/S = 1$) range is the same at about 3 km.

To calculate the range at which the constant gain system reverts back to a constant power system, we use (2.18) and (2.19) as follows:

$$P_j = P_i G_A$$

$$P_i = \frac{P_t G_t}{4\pi r^2} \cdot \frac{G_i \lambda^2}{4\pi}$$

Hence,

$$P_j = \frac{P_t G_t}{4\pi r^2} \cdot \frac{G_i \lambda^2}{4\pi} \cdot G_A$$

Thus,

$$r = \left[\frac{P_t G_t}{P_j G_j} \cdot \frac{\lambda^2}{(4\pi)^2} \cdot G \right]^{1/2} \tag{2.21}$$

Using the parameters in the examples, we show that the range at which the constant gain system reverts to a constant power system (begins to limit its J/S value), is at

$$r = 9 \text{ km}$$

This is the precise value as has been found graphically for the parameters used.

2.12 EFFECT OF PULSE COMPRESSION

Modern radars are being designed with intrapulse modulation of the transmitted radar signal for one or more of the following reasons:

1. To spread the power spectrum of the radar signal which minimizes the probability of detection by a hostile receiver.
2. To produce short pulses for fine target resolution even with the use of long pulses.
3. To produce high peak pulse powers (after compression) with the use of long pulse, high average power transmissions.

4. To decrease the J/S ratio of jammers with increased radar processing gain not available to the jamming signal.

The intrapulse modulation can take any coding form which is reproducible by the radar receiver on the reflected signal. Two important forms of modulation generally used are frequency modulation and phase modulation.

2.12.1 Pulse Compression Using Frequency Modulation

Frequency modulation within the pulse is often referred to as *chirp* because of the sweeping frequency program during the length of the pulse. This is shown in Figure 2.19. The length of the pulse depends on the compression gain and compressed pulsewidth desired. Typical values are 20 dB for the processing gain and 0.5 μs for the resultant pulsewidth. In this case, the transmitted pulse is typically 50 μs wide. It is during these 50 μs within the pulse that the sweep of the carrier frequency is accomplished, as shown in Figure 2.19.

Pulse compression of the chirped radar signal is as shown in Figure 2.20. As indicated, the delay of any part of the pulse is dependent on the carrier frequency at that point; the delay at the initial point of the pulse is maximum, while the delay at the end of the pulse is minimum. In essence, all parts of the long pulse arrive at the output of the delay device within the resultant 0.5 μs; compression of the pulse is thus accomplished. We should observe that the compressed pulse output is not observed until all of the 50 μs of pulse length has passed through the delay device. The net result is that the leading edge of the compressed pulse does not occur until approximately 50 μs after the leading edge of the incoming stretched pulse.

A long pulse which enters the delay device at a carrier frequency contained within the delay device but without the chirp modulation will not enjoy the pulse compression as does the radar signal. A short pulse (equal to the compressed pulse length) will exit the delay device in its original form with a delay defined by its carrier frequency. Note that this delay will almost always be less than the total delay experienced by the radar pulse. This suggests that an ECM pulse signal can be made to occur at the output of the delay device at a shorter range than the target reflecting the chirped radar signal. However, the ECM signal would need to compensate for the lack of processing gain enjoyed by the radar signal. This gain can be on the order of 20 dB.

The net result for an ECM system is a reduction in the effective J/S equal to the processing gain achieved with the pulse compression. An increase in the ERP of the jammer is required to overcome this loss, unless the interfering signal which has the matching characteristics of the reflected target signal can be generated. Because of the complex nature of the intrapulse modulations used in pulse compression radars, a true microwave storage system is required.

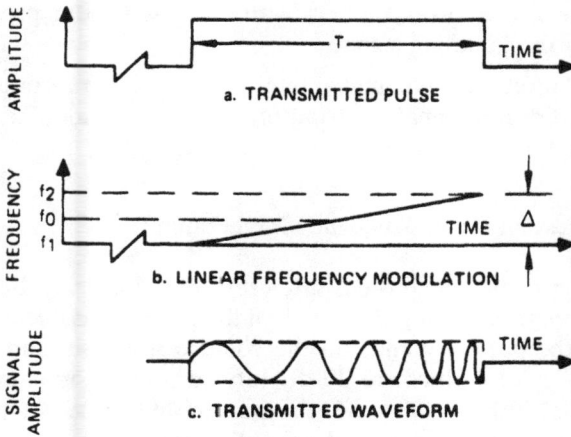

Figure 2.19 Transmitted waveform of a linear FM pulse (from [1]).

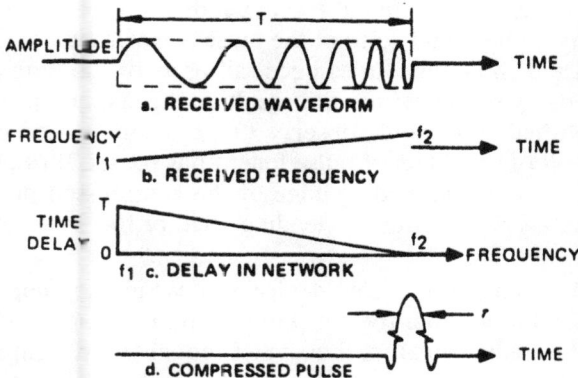

Figure 2.20 Pulse compression of a linear FM pulse (from [1]).

A saving feature for the ECM system is that the radar detection bandwidth must be compatible with the resultant short pulse rather than the transmitted long pulse. The net effect is that the radar receiver is subjected to all interfering signals which have a carrier frequency within the range of the *spread spectrum*.

2.12.2 Pulse Compression Using Phase Modulation

Figure 2.21 depicts the characteristics of a radar pulse when it is phase modulated to produce pulse compression of the reflected signal. As shown, a predetermined

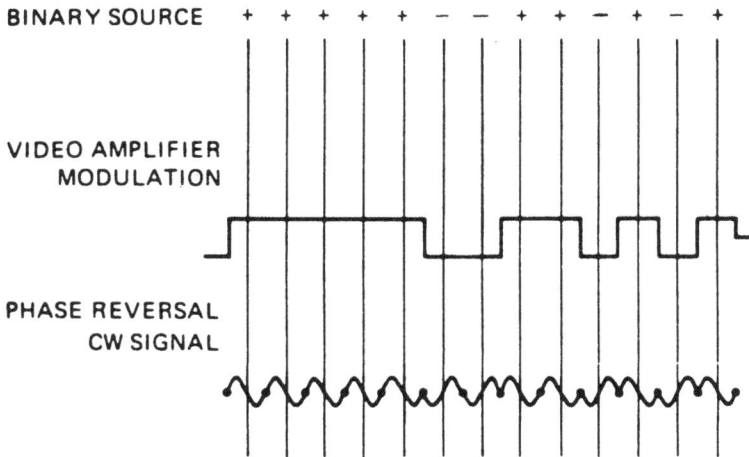

Figure 2.21 Waveform of typical phase shifted pulse (from [1]).

program of phase shifts of the carrier frequency is used to phase modulate the long radar pulse. For the reflected radar pulse, a network of phase shifters, as shown in Figure 2.22, is used to reconstruct the carrier frequency with demodulation of the phase shifts. Only when all of the long pulse is within the phase network, and only for a short period of time (the compressed pulse time), is a signal presented at the output of the network. Because of the phase matching network, only signals with the predetermined phase modulation enjoy the processing gain of this mechanism.

Note again that the output is not presented until the trailing edge of the uncompressed pulse. As with the chirp device, a pulse without the modulation will not enjoy the processing gain. However, unlike the FM pulse compression, a short pulse interfering signal anywhere within the stretched pulse will appear at the output at the same time as the compressed pulse. Unless the interfering signal energy level is increased to compensate for the lack of compression gain enjoyed by the true signal, it will have little effect on the radar detection process. Figure 2.23 shows an example of a decoded output waveform (using 13-element Barker code).

2.12.3 Low Probability of Intercept (LPI)

Although compression of the radar pulse produces a signal equivalent to one which would be produced by a high peak power transmitter, the actual transmitted signal is as much as 20 dB smaller in peak power, but much greater in average power

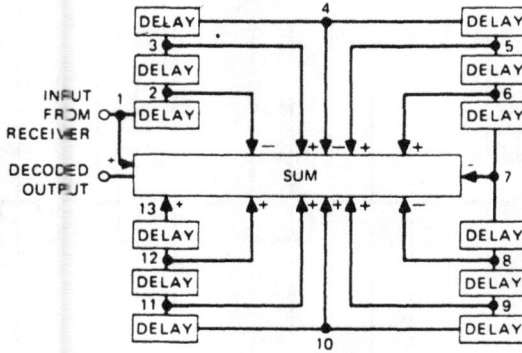

Figure 2.22 Pulse compression of phase shifted pulse (from [1]).

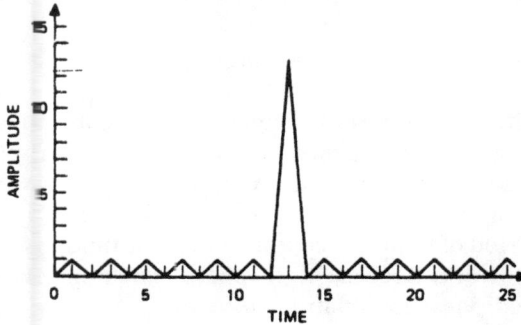

Figure 2.23 Compressed phase shifted pulse (from [1]).

for the same PRF. This is because the power spectrum of the stretched pulse is spread over a very wide frequency range due to the frequency or phase modulation. In many cases, this spread spectrum might be below the noise level of a hostile receiver which was intended to detect the presence of the radar signal, and would have if the real peak power transmitter was used.

Whereas LPI radars employ intrapulse modulation to spread the spectrum of the transmitted signal, they also transmit at a power level that is just adequate to yield a suitable signal at the radar receiver, yet undetectable by a hostile receiver which may be in the environment.

2.13 SUMMARY

In this chapter, formulas were developed which can be used to determine the radar signal power level, S, as well as the jamming signal power level, J, received at the radar antenna terminals. These formulas are then used to derive the (J/S) equations, which are needed to determine the ERP levels required by ECM transmitter systems. In addition, we showed how, with the use of logarithmic charts, the solutions to the equations can be represented by a series of straight lines, which are easily drawn with the calculation of only one or two points on the line. With the use of the straight line representations, an infinite number of solutions are available to designers or system engineers.

REFERENCE

1.   Hovanessian, S.A., *Radar System Design and Analysis*, Norwood, MA: Artech House, 1984.

Chapter 3
Search Radar Range Countermeasures

3.1 INTRODUCTION

Here and in Chapter 4, we will discuss the range and angle measurement functions of search radars. This discussion includes a description of the techniques used by the radar to measure the modulations of the reflected signal. These modulations are dependent on the position of the target vehicle relative to the radar. As indicated in Chapter 3, range position is determined by the time modulation of the radar transmitted signal. Angle is measured by noting the position of the radar antenna when detection of the reflected signal takes place; generally this is when the reflected signal power amplitude reaches a maximum as the antenna scans.

In addition, we discuss the various modulations that can be imposed on an ECM system's transmitted signal to confuse the radar range and angle measurement circuits as to the true position of the vehicle being protected by the ECM.

The intent of this chapter (or this book) is not to describe all of the details of radar range and angle measurement techniques. Our discussion of radar design is carried only far enough to allow appropriate discussion of ECM operation and its effect on these radar functions. For example, in this chapter there is a somewhat lengthy discussion of pulse repetition frequency (PRF) selection criteria and *second-time-around echoes* (STAE), primarily because these aspects of range measurement are extremely important for effective ECM operation and performance.

In these two chapters, all search radars are treated generically, so we assume that variations between different search radars are primarily in the volume of space searched; this factor has an effect on search speed and dwell time on target. For example, a surveillance radar used for early warning is responsible for "around-the-clock" (360° in azimuth) coverage about its position. Because its primary responsibility is to provide early warning detection (with secondary interest in target position accuracy), the dwell time on target is much less than that required in an acquisition radar, which is designed to make more accurate measurements of target parameters, and thus has a higher data rate or dwell time on target. Therefore, acquisition radars have lower search volume requirements and provide higher data rates.

Radars included in the category of *search* radars are EWRs, HF radars, and acquisition radars, as well as the acquisition function of radars that otherwise fall into the category of tracking radars.

Because the ECM techniques discussed in this book are generic in nature and are only secondarily affected by radar search volumes and dwell times on target, this generic approach is justified.

3.2 RADAR RANGE MEASUREMENT

As discussed in Chapter 1, all radars measure the distance (range) to a reflecting target by transmitting a signal and measuring the time elapsed from the instant of transmission to the instant of detection of the target reflection (skin return) in the radar receiver. Because the radar signal must travel out to the reflecting vehicle and then back to the radar, the travel distance (and thus the elapsed time) is twice the one-way distance (time) to the vehicle.
Therefore,

$$2r = ct$$

or

$$r = (1/2)\, ct$$

where

r = range to target
c = speed of radar wave travel (speed of light) = 3×10^8 m/s
t = time of travel from radar to target to radar

This equation is true regardless of which type of radar is used: pulsed, CW, FM/CW, *et cetera*. The only difference is the reference point used in the radar from which the travel time is measured. In the case of the pulsed radar, it is usually the leading edge of the transmitted pulse; in the case of the CW radar it can be a point (or phase) of the carrier wave; in the case of the FM/CW radar, it is usually a reference on the frequency modulation characteristic of the CW carrier. Each of these is discussed later in this section.

In order to show typical elapsed times involved in the solution of the above equation, let us calculate the elapsed time for a target at a range of 1 km:

$$t = \frac{2r}{c}$$

$$t = 2 \times \frac{10^3}{3 \times 10^8} = 6.7 \ \mu s/km$$

As shown, the total radar signal travel time for a reflector at one kilometer is approximately 6.7 µs, which translates to 670 µs for a target at 100 km. This illustrates a significant advantage of radar operation, that is, the ability to measure distant target parameters in an extremely small fraction of time. This, in addition to the radar's ability to observe targets at night, in rain, fog and smoke, makes the radar a very important instrument of warfare.

3.2.1 Pulse Radars

Figure 3.1 depicts the range measurement process in a pulsed radar. A pulse of a length on the order of microseconds is transmitted into space and reflected from a target within the space volume; the magnitude of the reflected power, as indicated in Chapter 2, is dependent on the effective RCS of the vehicle reflecting the signal. The reflected signal radiates in all directions and if the reflected signal power level (as intercepted by the radar) exceeds the detection threshold level, the echo pulse is displayed on the scope, as shown in the figure. The displacement in time is dependent on the distance to the reflecting vehicle. The timing circuits in the radar are calibrated to convert this time displacement to the distance to the reflecting vehicle. The scope can also be calibrated to indicate the target range directly.

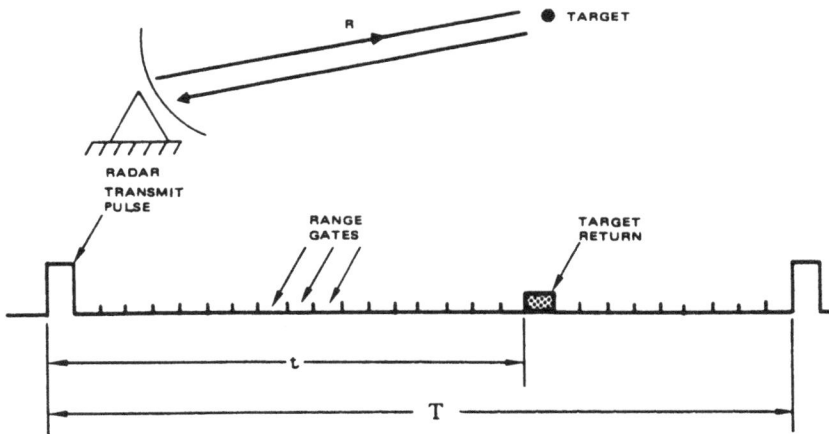

Figure 3.1 Pulse Radar Range Measurement.

Figure 3.1 shows an *A-scope presentation,* which is a display with a y-coordinate that represents the amplitude of the pulse received while the x-coordinate is a measure of time of travel from transmission of the pulse. The first pulse on the display is the transmitted pulse, representing the time at which the pulse is transmitted and timing is initiated. Figure 3.2 shows radar transmission types from a variety of radars.

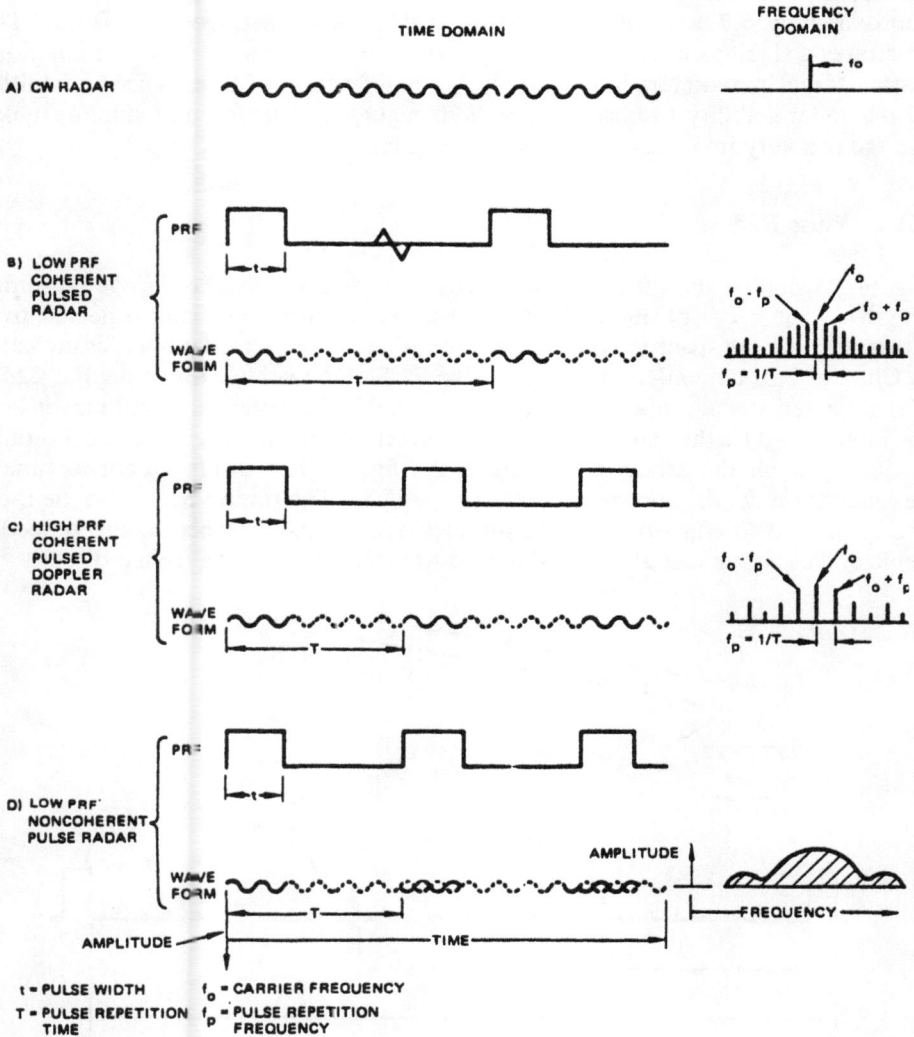

Figure 3.2 Types of radar transmissions (from [1]).

Figure 3.3 presents other forms of displays which are used to measure the time modulation of the reflected signal for range measurement. Figure 3.3(d) is the *plan position indicator* (PPI) display, which is most often used by search radars. The PPI display is in polar coordinates, with the angular parameter representing the angular position of the radar antenna and, thus, the *angle of arrival* (AOA) of signals received at that position. The radial displacement represents the round-trip of travel of the transmitted pulse to the target, and is calibrated to provide the measured range to the target which caused the reflection. In conventional pulsed radars, the target blip on the display is caused by the detected video from the reflected pulse (this is referred to as *raw video*). In advanced radars, which use signal processing techniques to remove noise or clutter, the target position is presented synthetically. This is because the raw video is lost in the noise (clutter) cancellation process, and cannot be presented on the display in "raw" form.

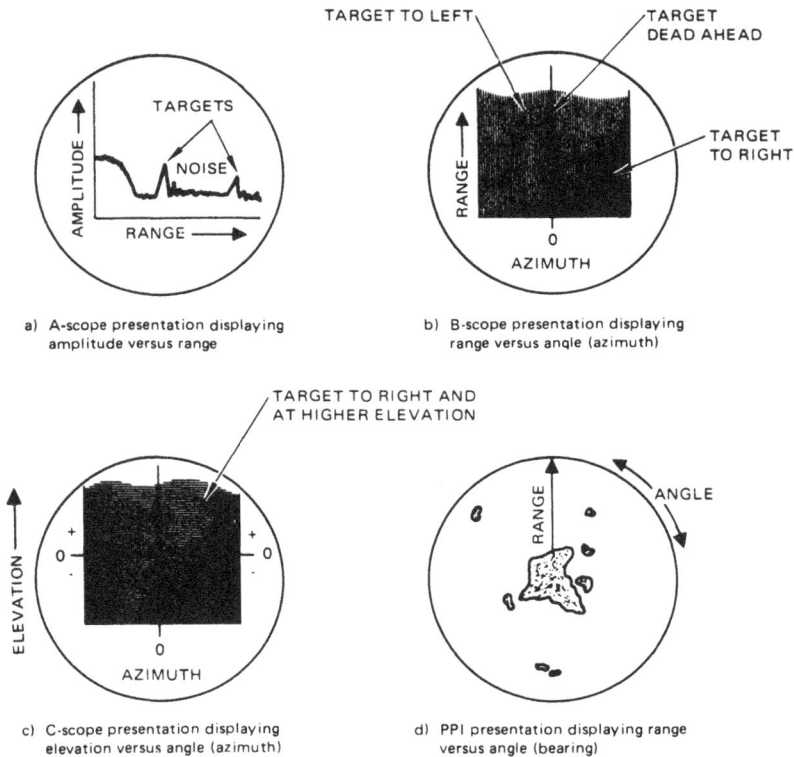

a) A-scope presentation displaying amplitude versus range

b) B-scope presentation displaying range versus angle (azimuth)

c) C-scope presentation displaying elevation versus angle (azimuth)

d) PPI presentation displaying range versus angle (bearing)

Figure 3.3 Common forms of radar displays (from [1]).

Figure 3.3(b) depicts a display which presents the same information in rectangular coordinates; the y-coordinate displays the reflected signal elapsed time (range) while the x-coordinate presents the angular position of the radar antenna This type of display is generally known as a *B-scope presentation*. As the figures show, the resolution of targets in angle is better in the B-scope presentation than in the PPI-scope presentation, especially at the inner ranges. Therefore, radars which are required to detect targets at ranges close to the radar are more likely to use the B-scope presentation.

Although there are other methods of presenting this information to an operator, they are not germane to the discussion in this book. As far as this chapter is concerned, three methods of presentation—the A-, PPI, and B-scopes—are adequate, and will be used to discuss the effects of the various ECM techniques designed to confuse search radars. Of the three, only the A-scope does not provide a measurement of target angle, only target range; it does, however, provide a measure of the amplitude of the detected signals, a parameter which is not readily available on the other two displays.

Radar detection circuits are also designed to measure the range to the targets automatically. As shown in Figure 3.4, a series of range gates can be arranged contiguously, with the first range position placed immediately at the end of the pulse transmission, and each subsequent gate contiguous to the preceding one. Generally, the width of the gate is equal to the radar pulse length. The range is determined by noting the position of the gate within which the return pulse is detected; these gate positions are calibrated to yield the true range to the reflecting vehicle. It is evident that if returns from targets at more than one range position are received simultaneously, target returns within the different gates are detected and measured uniquely. The detection circuits are the electronic equivalent of the A-scope display, and as such do not provide the angle measurement to the detected signals. Automatic target detection is shown in Figure 3.5.

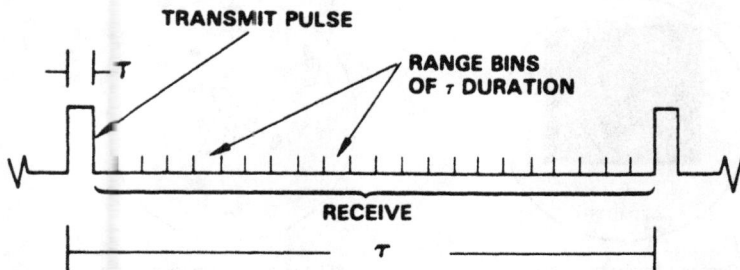

Figure 3.4 Range bins in pulse radar (from [1]).

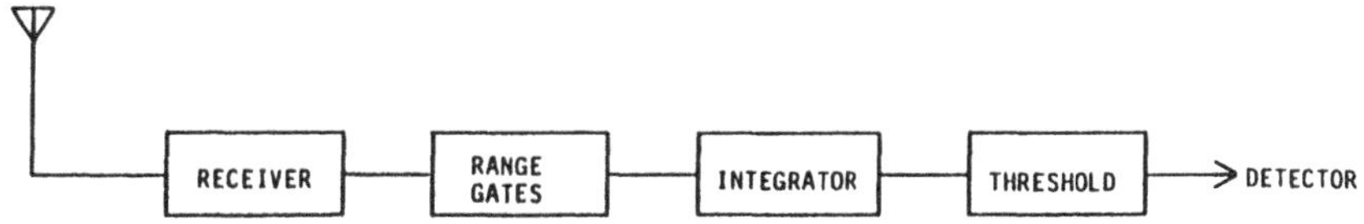

Figure 3.5 Automatic target detection.

A simplified block diagram of a conventional pulse radar is shown in Figure 3.6. As shown, a timing trigger is used to pulse the transmitter; this trigger sets the time counter to zero as well as the beginning of the sweep in the x-direction of the A-scope display shown in the figure. The *automatic frequency control* (AFC) circuits ensure proper tuning to the radar carrier frequency during all the time until the next pulse is transmitted, which may or may not be at exactly the same carrier frequency as the preceding pulse. All reflected signals which enter the antenna and exceed the detection threshold of the radar receiver are displayed on the same A-scope. The position of the received pulse (in the x-coordinate) is dependent on the time of travel of the reflected pulse (range to the target). As discussed previously, this scope can be calibrated to give the range measurement directly.

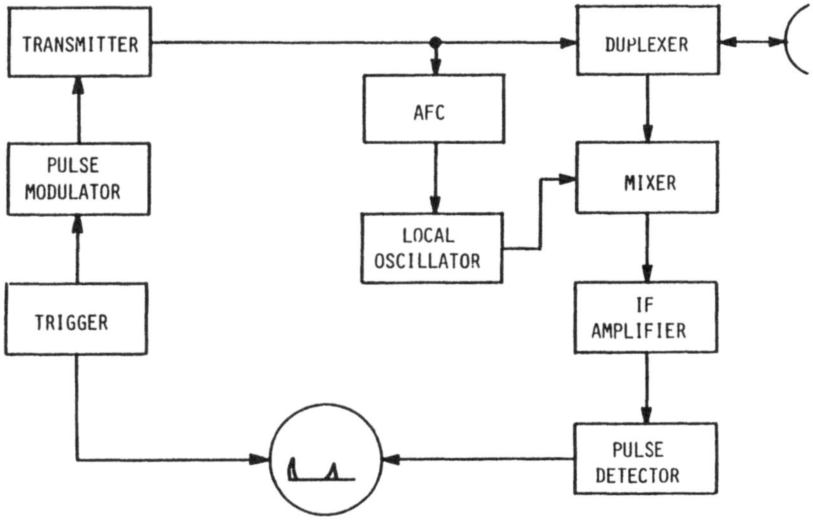

Figure 3.6 Conventional pulse radar diagram.

3.2.1.1 Second-Time-Around Echoes (STAE)

Conventional pulsed radars are designed to continue pulsed transmissions even when making measurements on a single target, because more data obtained on any one target makes the target detection and target position measurement more reliable. When multiple targets at different angles are expected, the radar must also continue transmitting to be able to detect targets at the various main beam positions of the searching radar.

Now the question arises as to which pulse repetition rate to use in the radar design. One of the limitations on the pulse repetition rate of a conventional non-coherent radar is the limitation in the duty cycle of the transmitter device (i.e., the magnetron, traveling wave tube (TWT), crossed field amplifier (CFA)). However, Figure 3.7 depicts another important limitation. As shown in the figure, a second pulse is transmitted before the reflection from the preceding transmitted pulse has entered the detection circuits so that the radar pulse travel time for the target is greater than the time interval between transmitted pulses. This pulse return is referred to as a STAE. When this occurs, the time measurement to the second-time-around pulse is based on the second transmitted pulse rather than the first (which caused the reflection); this is because each transmitted pulse is used to reset the signal time measurement. As shown, this range measurement is ambiguous; the measurement must be made from the first pulse for true, or unambiguous, range measurement.

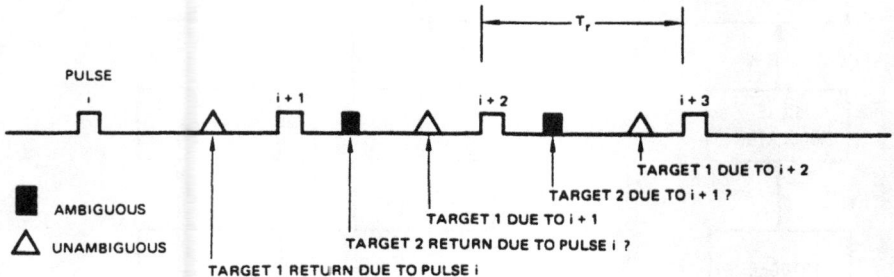

Figure 3.7 Ambiguous target returns (from [1]).

In order to prevent STAE, the PRF or, conversely, the time interval between transmitted pulses should be selected so that the time interval is long enough to allow the return of targets at the most distant range of interest before another pulse is transmitted. Therefore, if the pulse interval time = 1/PRF, we have:

$$t_{max} = 2\,r_{max}/c = 670\ \mu s$$

for $r_{max} = 100$ km
    $PRF = 1/t_{max} = 1/670 = 1492$ pulses/s

With this PRF, only returns from targets or other reflectors at ranges greater than 100 km will enter the radar receiver as STAE.

Radar designers have resorted to a number of techniques which eliminate the effect of STAE. One method is to change the carrier frequency on each pulse. As described previously, the receiver is tuned to the frequency of the signal at each transmission via its AFC circuits. Only radar echoes which return before the next pulse is transmitted are detected and measured by this type of radar, so STAE are no longer within the detection bandwidth of the retuned receiver.

In many cases, a pulsed radar is not able to change its carrier frequency on a pulse-to-pulse basis to eliminate STAE. In these cases, radar designers have resorted to the technique of varying their PRF in such a manner that the STAE can be recognized and discarded.

Figure 3.8 depicts a case of STAE; as shown, the received pulse R_4 is due to transmitted pulse T_4, R_5 due to T_5, R_6 due to T_6, *et cetera*. In each case, the received pulse is detected at the radar after the transmission of a succeeding pulse. In the case cited for a stable PRF, we can see that the ambiguous range interval (the time interval between the most recent transmitted pulse and the detected pulse), is constant. The result is that, although the interval and the range measurement are ambiguous, the pulses arrive at the same point in time relative to the nearest transmitted pulse. These pulses will produce a realistic blip on the

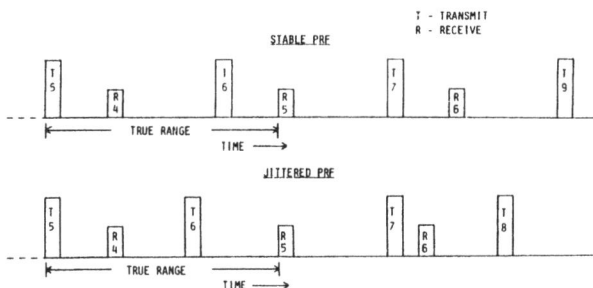

Figure 3.8 Second-time-around echoes (STAE).

displays, but at the ambiguous range. The detection circuits or operator are unable to determine that the range measurement is ambiguous. As shown in Figure 3.9(a), the pulse is detected at the same point in time and will serve to integrate in a range gate or scope display as if it is located at that unambiguous range position. This could be interpreted by an unsuspecting operator as the true position of a target.

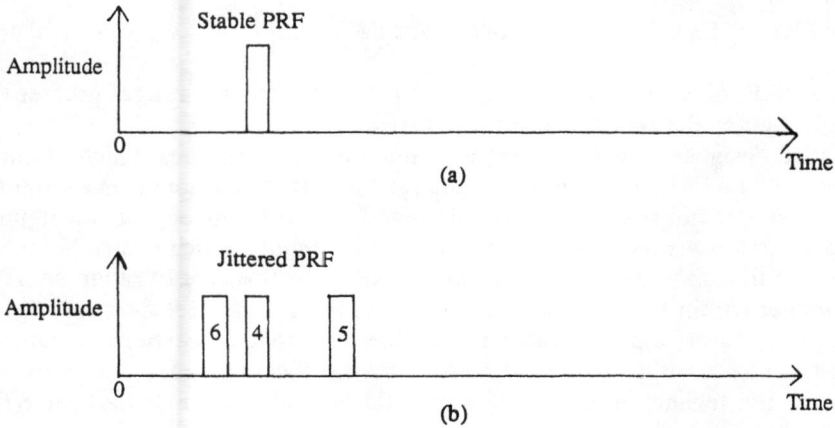

Figure 3.9 Effect of jit ered PRF.

On the other hand, if the PRF is varied in some manner, as shown in Figure 3.8(b), we can see that the ambiguous range interval varies from pulse to pulse, even though the unambiguous (true) range interval is constant for each pulse transmission.

Figure 3.9(b) shows the effect of *PRF jitter,* as this radar mode of operation is called. The detected STAE pulses do not occur at the same point in the time interval as they would if they were first-time around echoes. The net result is that there is no integration of pulses either in any range gate or any position on the scope. On any of the three displays, the STAE will appear smeared at the ambiguous range position, if it appears at all. The jitter is then easily identifiable by the operator or the electronic circuits as being due to a STAE.

3.2.2 FM/CW Radars

Before discussing FM/CW radars, we will introduce the basic CW radar, which is used to extract reflected signals from interfering clutter and noise by differentiating between moving targets and nonmoving targets. These radars are designed to measure the carrier frequency shift experienced due to the moving target when compared to the carrier frequency of the transmitted signal. Because these frequency shifts are extremely small (on the order of kilohertz) compared to the carrier frequency which is generally in the order of kilomegahertz), the radar must preserve the radar carrier frequency very accurately by using extremely stable and coherent transmitters.

Figure 3.10 depicts a simplified diagram of a CW radar. As shown, an extremely stable coherent transmitter is required, as well as an extremely stable local

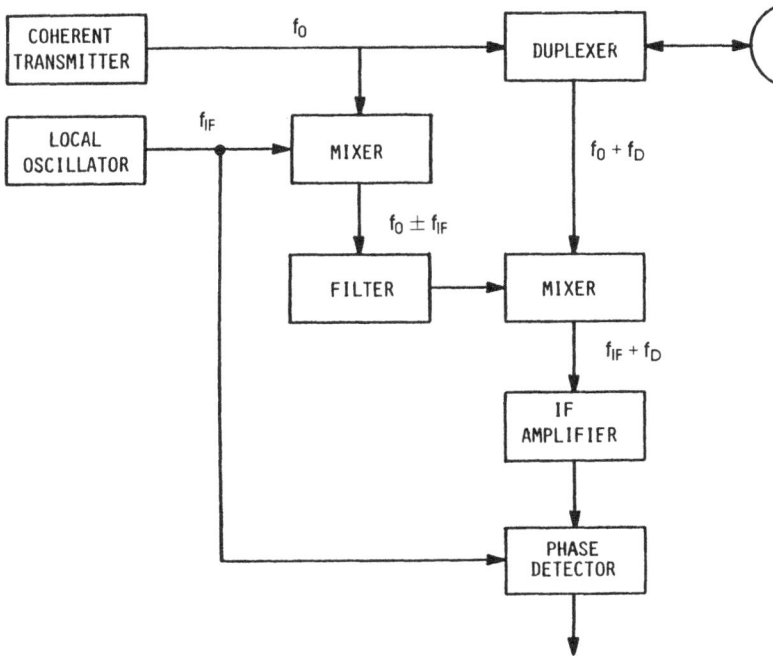

Figure 3.10 CW radar diagram.

oscillator (which is properly tuned to the transmitter frequency). As indicated, the return signal entering the radar receiver via the duplexer is heterodyned down to the intermediate frequency amplifier. The phase detector provides a measurement of the small frequency shift experienced due to the moving target reflecting the signal; a nonmoving target experiences no frequency shift. This frequency shift is known as the *doppler shift* and is directly proportional to the relative range rate between the target and the radar.

We may use Figure 3.11 to develop the amount of doppler shift experienced:

$$S_t = \sin(2\pi f_0 t)$$

$$S_r = \sin\left(2\pi f_0 t + \frac{2\pi R(t)}{\lambda}\right)$$

$$R(t) = \dot{R}t$$

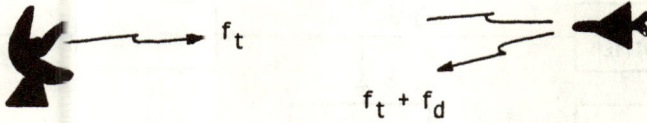

$$f_t = \sin(2\pi f_0 t)$$

$$f_t + f_d = \sin\left(2\pi f_0 t - \frac{2\pi R(t)}{\lambda}\right)$$

$$R(t) = \dot{R} t$$

$$f_t + f_d = \sin\left(2\pi f_0 t - \frac{2\pi \dot{R} t}{\lambda}\right)$$

$$f_d = \frac{\dot{R}}{\lambda}$$

Figure 3.11 Doppler effect.

$$S_r = \sin\left(2\pi f_0 t + \frac{2\pi \dot{R} t}{\lambda}\right)$$

Comparing the reflected frequency with the transmitted frequency,

$$f_d = \left(f_0 + \frac{\dot{R}}{\lambda}\right) - f_0 = \frac{\dot{R}}{\lambda} \tag{3.1}$$

If a coherent receiver is located aboard the vehicle and is exactly tuned to the radar transmitter frequency, the doppler shift measured would be the one-way doppler shift. The important point to remember about this calculation is that a CW radar signal impinging on the vehicle which is amplified for retransmission does contain the one-way doppler shift, so that the amplified and retransmitted signal (as received by the radar) will contain the two-way doppler shift as it would in a reflection from the vehicle. Therefore, any change in the frequency of the signal as it passes through the amplifying system on the vehicle is a change in the doppler frequency as detected at the radar receiver. For example, a 5 kHz frequency shift applied in the amplifying system is detected as a 5 kHz shift in the radar detected doppler frequency as received from the target reflection.

Using practical dimensions for the equations, we can show that the two-way doppler shift is equal to:

$$f_d = \frac{2\dot{R}}{\lambda}$$

$$f_d(\text{Hz}) = \frac{55(\text{km/hr})}{\lambda(\text{cm})} \tag{3.2}$$

and at 10 GHz where $\lambda = 3$ cm; f_d in Hz $= 18 \times V$ in km/hr.

It is not readily possible to measure the time of travel of the transmitted signal with a pure CW radar because there is no reference on the transmitted signal that can be conveniently used as a reference for the return signal. We may suggest that the phase shift on the reflected signal can be measured, except that, for the carrier frequencies of interest, one cycle of the wave (360° phase shift) is on the order of 3 cm, which means that a 360° phase shift is experienced for every 3 cm change in (two-way) range. Although the radar can measure the fractional change in phase, it is not practically possible to determine the number of full cycles of phase shift experienced at any range position.

Therefore, in order to use a CW radar for range measurement, it is necessary to impose some type of modulation on the transmitted signal which will be used as the reference. Frequency modulation of the carrier frequency is most often used. A typical modulation is shown in Figure 3.12. Shown as the dotted line in the figure, the return signal is shifted in time depending on the distance traveled by the transmitted wave. A point of reference is selected on the modulation, and is then used to determine the time of arrival of the same reference on the return signal. The figure shows that for this type of modulation, we can also measure the frequency difference between the two modulations, which is dependent on the range to the reflecting vehicle. Although other types of frequency modulation to provide an adequate reference for range measurement with a CW radar are possible, the FM characteristic illustrated is adequate to discuss the point being made here. A block diagram of an FM/CW radar is shown in Figure 3.13.

3.2.3 Pulsed Doppler Radars

Another method of modulation of an otherwise CW signal is amplitude modulation of the carrier, in which pulse modulation, as is done in a conventional pulsed radar, is employed (Figure 3.14). Pulse modulation provides the same degree of resolution of the range measurement as does a conventional pulse radar, but the pulsed doppler radar experiences extreme range ambiguity problems when high PRFs are used in these radars. A high PRF is often required to avoid ambiguities in the doppler (velocity) measurement.

Figure 3.15 illustrates the power spectrum of a CW radar signal as a function of frequency. The vertical line represents the power level of the transmitted CW

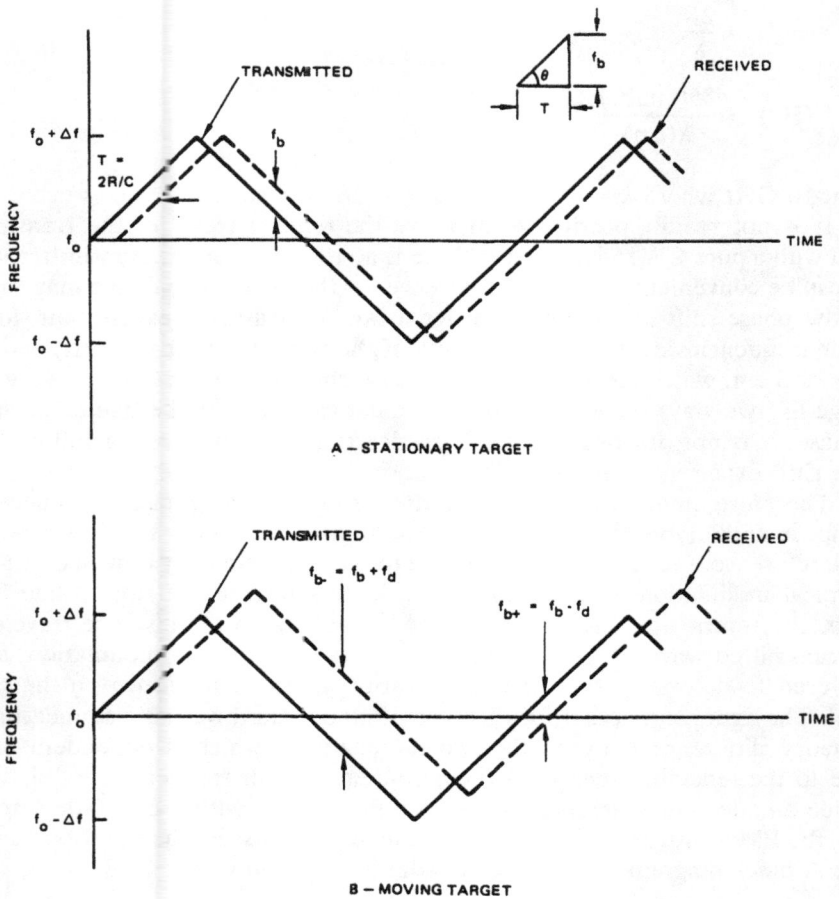

Figure 3.12 Range and range rate measurement in CW radars (from [1]).

signal at the carrier frequency of the radar. The power of the reflected signal from a moving target is shown displaced in frequency; the amount of the displacement is a function of the relative velocity between the radar and the moving target. A positive frequency shift is due to a closing target; an opening target at the same relative velocity would be at an equal distance to the left of the CW line.

Figure 3.16 illustrates the power spectrum of a pulse modulated carrier. According to Fourier analysis, the result is a line spectrum, the lines of which are distant from each other by an amount equal to the PRF. Analysis also shows that a moving target reflecting this transmission will produce target lines corresponding to each of the PRF lines in the spectrum.

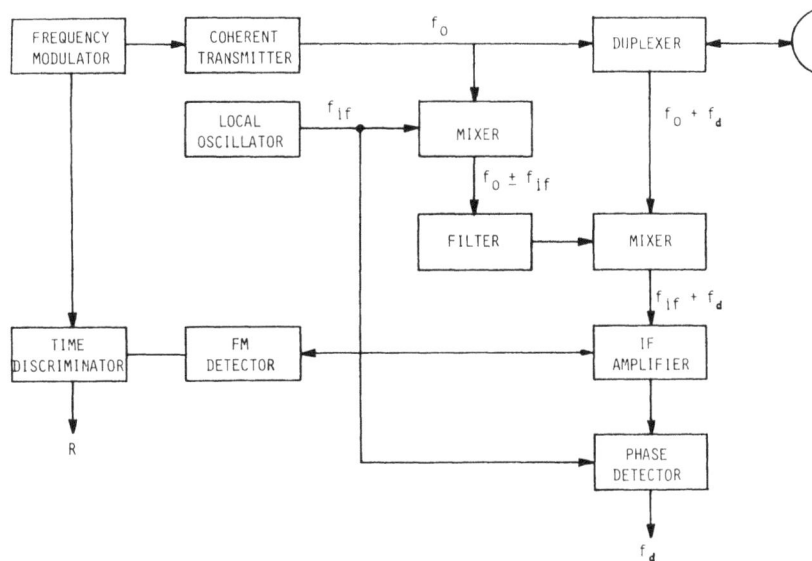

Figure 3.13 FM/CW radar diagram.

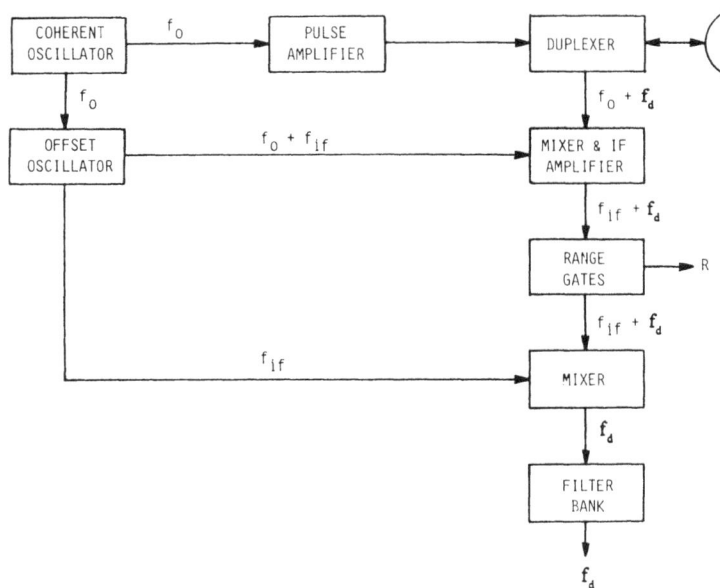

Figure 3.14 Pulsed doppler radar diagram.

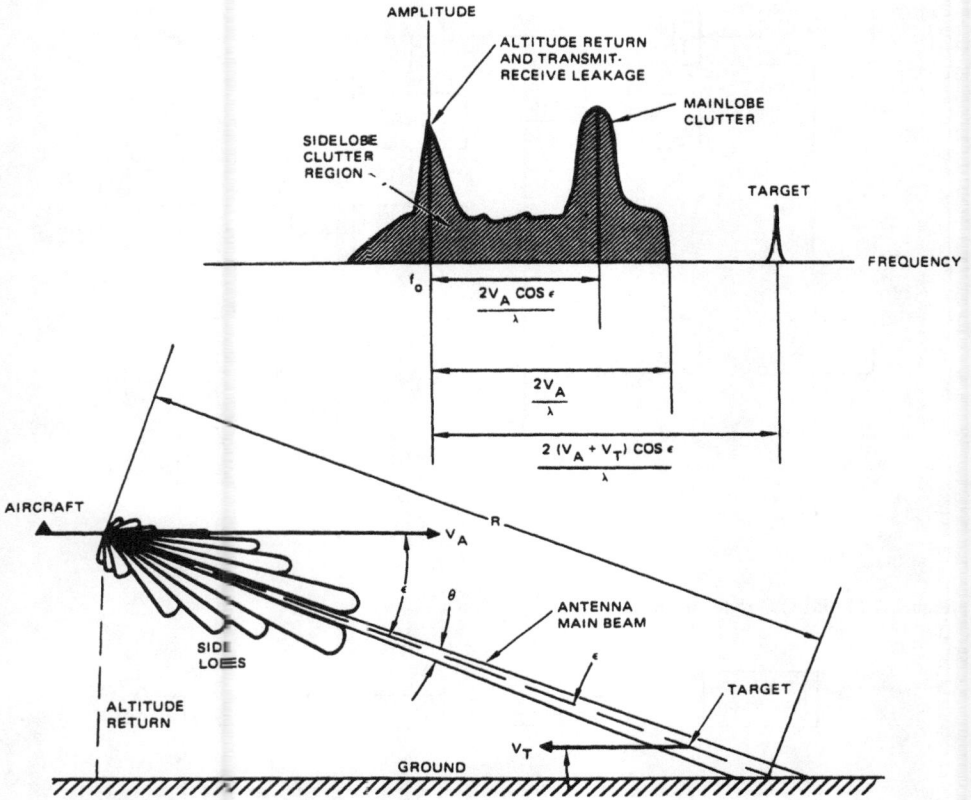

Figure 3.15 Signal and clutter spectrum of CW radar (from [1]).

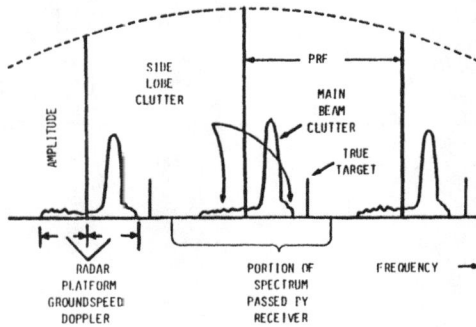

Figure 3.16 Pulsed doppler signal spectrum.

We can see that if the doppler frequency shift of a moving target is greater than the PRF, the target line shifts beyond the next PRF line, which results in an ambiguity in the doppler shift measurement. This ambiguity results because there is no immediate means of determining to which PRF line the target doppler shift is to be measured.

Furthermore, if the doppler shift is greater than half the distance between the PRF lines, an ambiguity exists as to whether the target line is due to an opening target (which is referenced to the second line) or a closing target (which is referenced to the first line). This effect dictates that, if the radar is to avoid velocity (doppler shift) ambiguities, the PRF selected should be at least double the maximum doppler shift expected, as follows:

$$\text{PRF} \geq 2 f_d \text{ (max)}$$

For $V = 500$ km/hr at 10 GHz,

$$f_d = 9000 \text{ Hz}$$

For a typical case, the minimum PRF is then

$$\text{PRF} = 2f_d = 18 \text{ kHz}$$

For this PRF, we can calculate the unambiguous range:

$$\frac{1}{\text{PRF}} = 55.5 \ \mu s$$

$$r = \frac{1}{\text{PRF}} \times \frac{1 \text{ km}}{6.7 \ \mu s} = 8.3 \text{ km}$$

For this case, a target at a range greater than 8.3 kilometers will be detected as a STAE. A target at 83 kilometers will enter the receiver as a *multiple-time-around echo* (MTAE) because as many as ten more transmitted pulses will have been transmitted before a reflected signal from any one of the transmissions will enter the radar receiver. A low PRF could be used to prevent multiple or even STAE, but this would result in severe velocity ambiguity problems. For example, with a PRF of 1 kHz, a target with a doppler shift of 10 kHz would fall 10 spectrum lines distant from the line to which it should be measured. In this case, a technique of velocity ambiguity resolution is needed. This mode of operation (low PRF) of a doppler radar is often used after the true target doppler shift has been determined with a high PRF mode, so that the ambiguity can be tolerated.

The net result is that the radar designer must choose between designing for elimination of range ambiguity or velocity ambiguity. In some cases, the radar design can tolerate both range and velocity ambiguities. In any case, the radar design must include methods of resolving these ambiguities either in range or velocity or both.

3.2.3.1 Range Ambiguity Resolution

Pulsed doppler radars are often operated at very high PRF rates, and, as shown in the previous paragraphs, are subjected to MTAE and a serious degree of range ambiguity. Even with high PRF rates, the radar provides very accurate measurement of the echo's position between transmitted pulses (ambiguous range). However, it is not known to which transmitted pulse time the echo should be measured for true unambiguous range determination.

This is akin to a clock whose hour hand is missing. On such a clock it is easy to determine what the time is within the hour, but not at what hour. So with a high PRF radar system, it is easy to measure when the pulse arrived within a pulse interval time, but a question exists as to which of the preceding pulses caused the reflection.

Radar designers have resorted to various techniques of resolving this ambiguity, the more popular being the application of the *Chinese remainder theorem*. (See Appendix C.) In the case of two or more timers, it is possible to determine where in the unambiguous period the instant of time of interest is located when measurements are made of the ambiguous interval in each of the timers, if the timers are mathematically related and periodically come into synchronism at a predetermined unambiguous time interval.

For example, in Figure 3.17, if the radar uses three PRFs so that the transmitted pulses of all three occur in synchronism periodically (as shown, this is at $t = 0$ and $t = T$), and these PRFs are lined up as shown, only at one position in the total unambiguous period do the return pulses line up. This occurs at the true unambiguous position relative to $t = 0$. We can see in the figure that the ambiguous time interval (time within the pulse interval) for each of the three PRFs is different from the others. However, by measuring this ambiguous time interval in each of the PRFs, mathematically determining the unambiguous time position is possible. Figure 3.18 shows a two-PRF ranging system.

We can see the same result with the three clocks in Figure 3.19. Each of the clocks is calibrated for 24 hours, with only the first clock requiring 24 hours for the single hand to make one complete revolution; the second clock requires 48 hours to make one complete revolution and the third clock requires 72 hours. If all three clocks are started at zero at $t = $ zero, the three hands will not arrive simultaneously at zero again until $(1 \times 2 \times 3) \times 24$ hours, or six days later. At

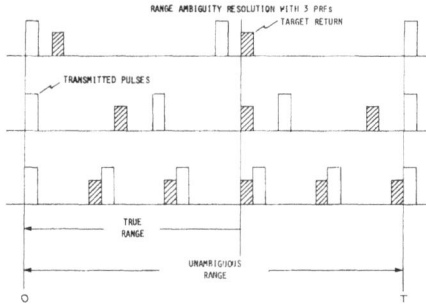

Figure 3.17 Transmit and receiver pulses in pulsed doppler radar (from [1]).

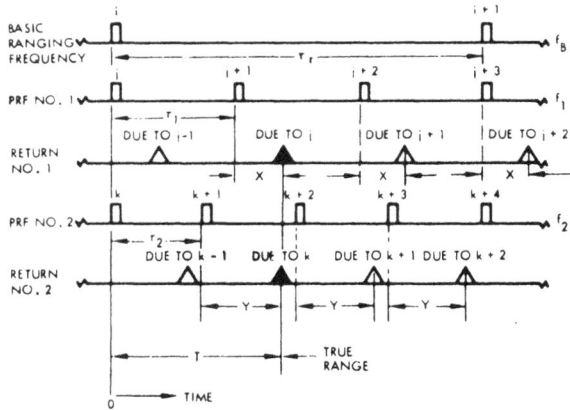

(▲ unambiguous, △ ambiguous)

Figure 3.18 Two-PRF ranging system (from [1]).

the end of the first day, the first clock will be at zero, but the second will only be at 12, and the third at 8. The net result is that for any instant of time within the six-day interval, a unique set of hand positions will define that instant of time in the unambiguous time interval. Table 3.1 shows the hand positions (ambiguous time positions) for a limited number of instant unambiguous times in the six-day period. We can see with this limited set that a unique combination of ambiguous hand positions exists for each unambiguous time position.

This is also true for the three-PRF radar method described above. The unambiguous time interval for true range to the target is the time between the periods

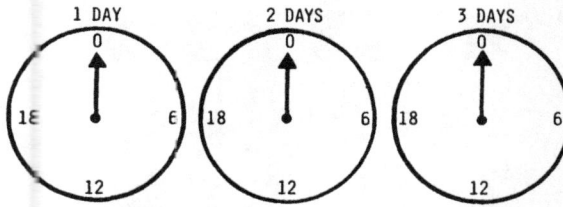

- POINTERS REACH ZERO SIMULTANEOUSLY ONLY ONCE EVERY 6 DAYS.

- POINTERS REACH A UNIQUE COMBINATION OF POSITIONS ONLY ONCE EVERY 6 DAYS.

- ANY ONE COMBINATION OF POINTER POSITIONS DEFINES A UNIQUE TIME WITHIN THE 6-DAY PERIOD.

Figure 3.19 Clock analogy for ambiguity resolution.

Table 3.1

Day:Hour	T1	T2	T3
0:0	0	0	0
0:6	6	3	2
0:12	12	6	4
0:18	18	9	6
1:0	0	12	8
1:6	6	15	10
1:12	12	18	12
1:18	18	21	14
2:0	0	0	16
2:6	6	3	18
2:12	12	6	20
2:18	18	9	22
3:0	0	12	0
3:6	6	15	2
3:12	12	18	4
3:18	18	21	6
4:0	0	0	8
4:6	6	3	10
4:12	12	6	12
4:18	18	9	14
5:0	0	12	16
5:6	6	15	18
5:12	12	18	20
5:18	18	21	22
6:0	0	0	0

when the transmitted pulses are synchronous. Therefore, each combination of ambiguous time interval measurements uniquely defines a time position within the unambiguous time interval. The interested reader is encouraged to consult the relevant literature to learn more about the methods of resolution of range and velocity ambiguities.

Although this method of resolving the range ambiguities associated with the operation of a radar with very high PRFs is practical when only one target appears within the ambiguous range intervals, the technique generates 'ghost' solutions when two or more targets are present within the same ambiguous pulse interval times. We can show that the number of solutions using this method is equal to the number of pulses occurring within the pulse interval raised to a power equal to the number of PRFs used, as follows:

Number of Solutions $= (n)^k$

where

n = number of targets within pulse interval;
k = number of PRFs used.

For example, if four pulses appear within the pulse interval, with a three-PRF system, 4^3, or 64 solutions result, with only four being true solutions.

This suggests a vulnerability in pulsed doppler radar operation which can be exploited by ECM designers. The reader must realize, however, that, for example, if the radar can differentiate between the multiple pulses within the pulse interval on the basis of doppler frequency, it can separate these signals before the range ambiguity solution is attempted.

As we have said, the intent of this book is not to discuss radar design details to any greater degree than is necessary to understand the ECM techniques and why they may or may not be effective. The discussion on the range ambiguity problem with pulsed doppler radars was presented to identify a serious vulnerability in this type of radar that can easily be exploited by ECM system designers. Therefore, the mechanization used to provide the range ambiguity solution is not presented here. The interested reader may refer to the literature for the design details.

3.3 RADAR RANGE ECM

This section describes the various methods used by ECM system designers to deny the hostile radar its capability to measure the range parameters of its intended targets. As discussed above, a radar relies on the measurement of the time of travel of its transmitted signal from the radar to the target and back to the radar.

ECM techniques are designed to provide interfering signals which are within the radar detection bandwidth and contain modulations which conceal the true position of the protected vehicle or confuse the radar as to the true location of its target.

The techniques discussed in this chapter are designed primarily to confuse or deceive the range measurement function of the victim radars. By themselves, range jamming techniques may or may not affect the angle measurement function of these radars. However, when used in combination with the angle jamming functions described in the next chapter, the ECM system can be very effective in aborting the mission of the defense system.

We must emphasize that it is absolutely necessary for ECM designers to have an in-depth understanding of radar design to design effective ECM systems. Readers interested in ECM system design are encouraged to study radar design principles and to keep abreast of new radar developments.

3.3.1 Target Cover

This section discusses ECM techniques which are designed to deny the victim radar its target range measurement capability by covering or suppressing the true echo return before it can be detected at the radar receiver. With these techniques, it is not readily possible for the radar to detect the true time of arrival of the reflected signal or any of the modulations which identify its position in range.

3.3.1.1 Noise

The more commonly used method of covering the true target signal is with a noise transmission designed to enter the radar receiver at, or near, the same time as the target reflected signal and (desirably) for a suitable period before and after the arrival of the pulse at the radar receiver. Transmission of noise during arrival of the pulse at the radar is a formidable requirement for ECM systems. This is because, to perform this function, the ECM system receiver must be able to measure or predict the time of arrival of the pulse at the target vehicle and be able to measure the carrier frequency of the intercepted pulse within nanoseconds of the leading edge of the intercepted signal. Although interception of the radar signal at the ECM receiver is almost simultaneous (within nanoseconds) with the radar reflection from the vehicle, any delay in the ECM system's transmission of a jamming signal will cause the ECM signal to arrive at the radar with a corresponding delay relative to the target echo.

Furthermore, assuming that the ECM receiver has the required frequency measurement capability, the ECM transmitter must be set for frequency and transmitting before the trailing edge of the intercepted signal passes through the receiver system. Otherwise, the ECM signal will enter the radar receiver after the true echo (with its proper position modulations) has been detected by the radar.

A method used by ECM designers to compensate for this problem is to continue transmitting at the frequency detected on each pulse for a period which encompasses several repetition periods of the radar. The success of this technique is based on the assumption that the radar transmitter has not changed its carrier frequency and is still receptive to signals at the frequency of the carrier measured by the ECM receiver. In this case, the signals generated by the ECM are accepted by the radar receiver at all times during the interpulse period, both before and after the arrival of the true target signal.

Even if we assume that the ECM receiver has the capability of transmitting at the correct frequency instantaneously with the arrival of the signal at the ECM receiver, continuous transmission of that carrier for several periods will produce coverage at shorter ranges from the radar to the target only if the radar continues receiving at that carrier frequency.

Modern pulse radar system designers have resorted to *frequency agility techniques,* which allow the radar to change its carrier frequency of operation in a rapid fashion, even on a pulse-to-pulse basis. This technique is used primarily to deny ECM systems the ability to produce interfering signals to arrive at the radar receiver before and during the expected reflected pulse for the reasons discussed above.

Carrier frequency agility is very effective, even against the optimum ECM receiver that can measure and set the ECM transmitter onto the frequency on an instantaneous basis. Because the radar receiver is retuned in frequency on a pulse-to-pulse basis, the ECM signal, which is not updated in frequency from pulse to pulse, will be rejected by the radar receiver for all times except for the period between the arrival of the target echo and the transmission of the next pulse. In this case, the interfering signal will never precede the true target position. Figures 3.20 and 3.21 display the B-scope and PPI presentation with a nonagile radar and the effect of frequency agility on ECM operation. As we can see in each of the displays, no interference exists at ranges of less than the true target range for the agile radar.

In order to counter the carrier frequency agility in advanced radars, the ECM system must resort to relatively broad frequency signal transmission, wherein the power spectrum of the transmitted signal contains all frequencies to which the radar may be tuned. Rather than attempt to measure and set its transmitter onto each radar pulse, the ECM transmitter system radiates simultaneously over the complete frequency range over which the radar can be tuned. This is accomplished with modern ECM receivers which monitor the frequency of the radar transmissions and establish their operating range. As shown in Chapter 2, this does result in a demand for a higher effective radiated power from the ECM transmitter due to the dilution of power over a bandwidth much greater than the detection bandwidth of the radar. This is a compromise which must be accepted if the transmitter is required to generate interference which precedes the time of arrival of the target echo at the radar receiver.

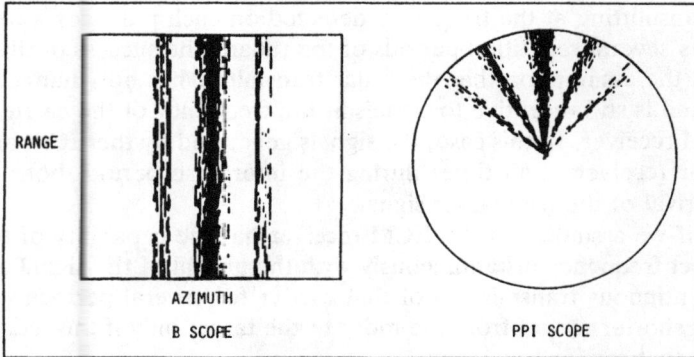

Figure 3.20 Noise jamming of nonagile radar.

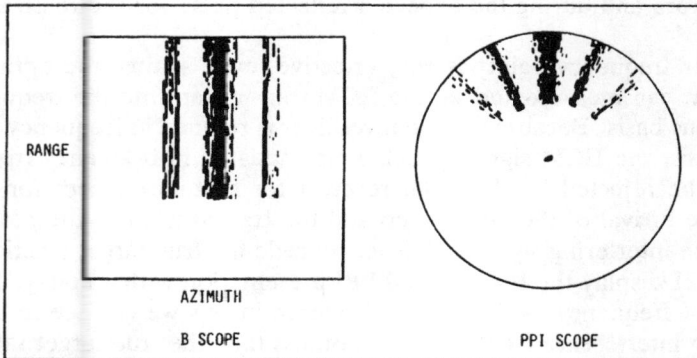

Figure 3.21 Noise jamming of agile radar.

Typically, the J/S requirement for effective coverage of the true target signal is on the order of zero dB ($J/S = 1$) (the interfering signal is equal to the reflected signal as it enters the radar receiver). Although this is an acceptable rule of thumb, the required J/S can vary as much as 10 dB on either side of zero depending on the detection requirements and the sophistication of the radar receiver circuits. Identification of the vehicle and its dynamics by the radar will require a higher signal-to-interference ratio than is required only for detection of the target's presence. Post detection processing and integration can improve the radar's signal-to-interference ratio by as much as 20 dB. One such effect was discussed with the pulse compression radar in Chapter 2.

The noise type of interference is effective against all types of radars (i.e., pulse, FMCW or pulsed doppler), the only difference being in the amount of J/S required. We must emphasize that pulsed doppler radars are required to maintain a very stable carrier frequency over many pulse interval periods because of the requirement to extract small frequency shifts due to motion of its targets. This requirement dictates that the pulsed doppler radar transmit exactly on the same frequency for an extended period of time to enable proper extraction of the doppler shift due to moving targets. Therefore, frequency agility on a pulse-to-pulse basis is not possible for pulsed doppler radars. This requirement makes them vulnerable to ECM systems which are unable to measure and set onto the radar frequency on a pulse-to-pulse basis.

3.3.1.2 Target Removal

Although there is no practical method of cancelling the reflected signal before it enters the radar antenna, much effort is being expended to devise techniques to prevent the detection circuits from displaying the target data on any of the scopes available to the operator or to be detectable by electronic circuits. Because of the requirement for subclutter target detection and the use of synthetic displays associated with the new radars, more sophisticated algorithms are being used to detect and analyze radar signals before presentation on operator displays.

These algorithms include examination of power levels of potentially interfering signals at or near the suspected target position, as well as neighboring signals in range and velocity. ECM signals interfering with proper operation of any of these algorithms can confuse the radar analysis circuits to the point that no target is recognized or displayed. The net result is elimination of the target signal from detection by the operator. Although removal of target from the display is certainly more effective than attempting to cover it with interfering signals, the techniques used to accomplish the desired result are highly dependent on *a priori* or predictable knowledge as to the algorithms used in the radar analysis circuits. This requirement makes these techniques extremely fragile and can be countered with relatively small changes in the radar detection logic.

3.3.2 Target Confusion

The target confusion technique suggested in this paragraph is more effective when considering ECM operation against the angle measurement mode of the search radars, and so will be discussed in more detail later. However, it is introduced here because it can also be used for confusion in the radar as to the proper range of its targets in its main beam.

3.3.2.1 Conventional Pulsed Radars

The intent of this technique is not to deny the radar the true reflected signal nor its time of arrival at the radar, which provides the true range measurement to the target. The intent of the target confusion technique is to present many targets to the radar at ranges different from that of the true target so that the radar is unable to differentiate between the true and false targets. Therefore, even though the true target is included within the multiple targets presented, the radar may not have the capability to identify it, and must then accept all target measurements as potential target positions. The PPI display in the Figure 3.22 depicts the presentations resulting from such an ECM system.

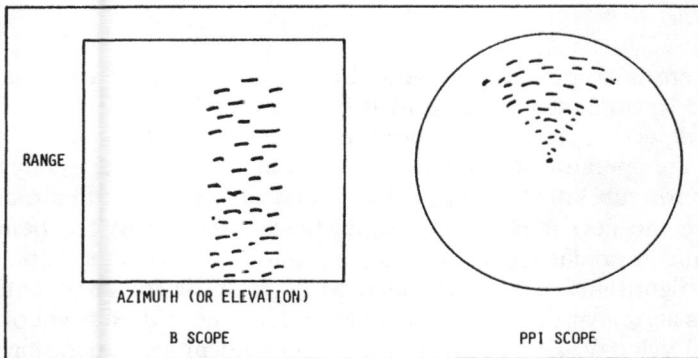

RANGE

AZIMUTH (OR ELEVATION)

B SCOPE   PPI SCOPE

Figure 3.22 False targets on nonagile radar.

The multiple target transmission system is subject to the same restrictions presented above for noise jamming of the pulsed radars. In particular, to be able to present realistic targets at ranges nearer to the radar than the true target (nearer targets), the ECM system must depend on frequency stability of the radar carrier frequency over at least two pulse intervals. As indicated above, this is an inherent requirement for doppler radars, but not for conventional pulsed radars. Therefore, in the case of a frequency agile radar, the pulses transmitted by the ECM and arriving at the radar after the radar has changed its frequency will not appear at inner ranges and the result will appear as in Figure 3.23. Again, the ECM system can resort to the use of a broadband transmitter (as suggested above for the noise jammer) but with the requirement for a higher ERP from the ECM transmitter because of the resulting power dilution.

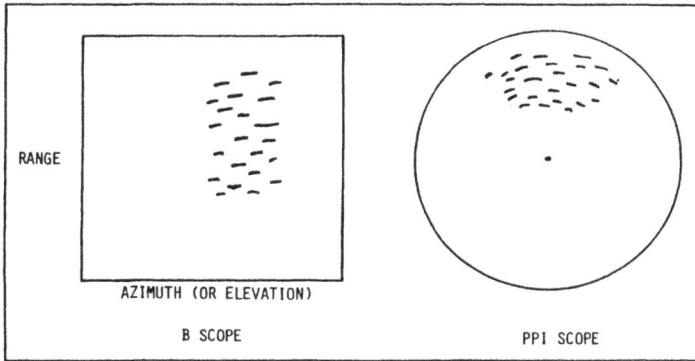

Figure 3.23 False targets on agile radar.

3.3.2.2. Pulsed Doppler Radars

A pulsed doppler radar is especially vulnerable to ECM techniques because of its inherent requirement for extreme carrier frequency and PRF stability. In addition, pulsed doppler radars operating at medium-high or high PRFs suffer from severe MTAE and subsequent range ambiguity resolution requirements. As indicated above, the solution to the range ambiguity problem can yield multiple erroneous range solutions when two or more targets are detected within the same ambiguous range interval. Although the multiple targets within the same range interval must be at the same angle as well as the same doppler to cause such multiple solutions, an ECM system can easily be constructed to meet these requirements.

Such a system is shown in Figure 3.24. As indicated, the signal entering the ECM system's receiving antenna is an exact duplicate of the signal reflected off the vehicle being protected by the ECM. If amplified and retransmitted without modulation, the signal will appear at the same range, angle, and doppler as the reflected signal. However, if this signal is stored in the coherent memory and exact duplicates are retransmitted at later times during the ambiguous pulse interval, these pulses will appear as shown in Figure 3.25.

The result is a number of targets (including the true target) within the same ambiguous range interval, which is the condition for multiple range solutions in the range ambiguity resolver. A total of four targets within the range interval with a three-PRF system will result in 64 solutions, only one of which is correct. We must point out that it is not required nor will it degrade the performance of the ECM technique if the multiple target ECM transmission overlaps into the next pulse interval.

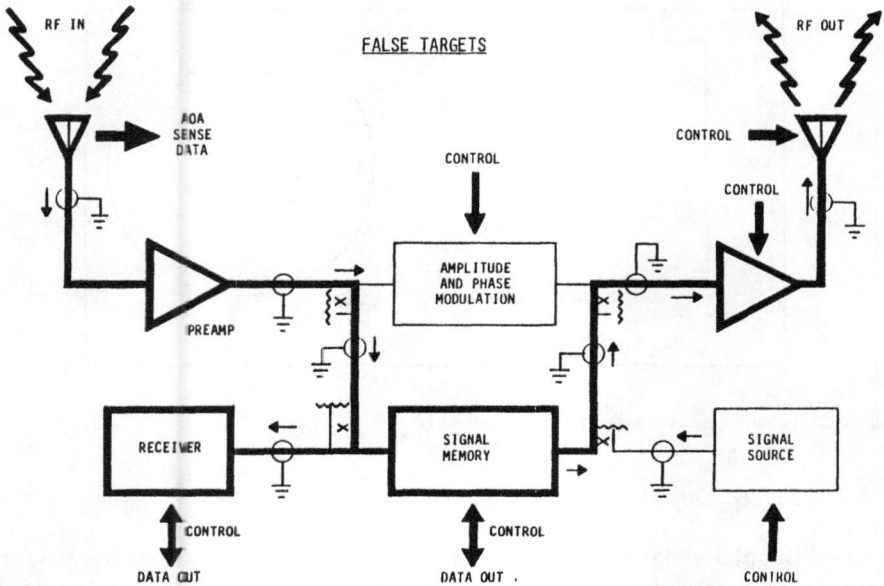

Figure 3.24 False target jamming configuration.

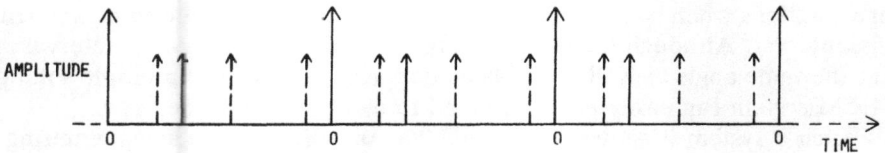

Figure 3.25 Jamming of range ambiguity resolver.

3.4 VELOCITY ECM

Conventional pulsed radars do not have nor do they need a requirement for instantaneous measurement of the range rate of their targets. If that parameter is required in a radar's mission, it is obtained by taking the time derivative of consecutive measurements of range. However, because measurement of the doppler shift of the received signal by a doppler radar is an instantaneous measure of relative velocity between the radar and the target, this parameter is the primary sorting parameter in doppler radars. This parameter is excellent in the process of filtering of targets in the presence of nonmoving target clutter. Measurement of the doppler shift immediately provides the radar a true measure of the range rate (velocity) of the target relative to the radar position.

Many doppler radars use PPI and B scopes to present the relative range rate (rather than range) as a function of angle to be the primary detection mechanism. This is primarily because the measurement of range requires more dwell time as well as computation time. In search radars, detection of target and angle are primary requirements; range rate measurements are generally of secondary interest.

Because doppler search radars use doppler filter banks or their equivalent in the detection and sorting of reflected signals, an ECM system can confuse the radar detection process by injecting a number of other signals which represent targets at other doppler frequencies into the radar receiver.

The same ECM system suggested previously for the range ambiguity ECM can be used by phase modulating either the amplifier or the coherent microwave signal storage system. In the former case, targets at the same range but different dopplers are generated; in the latter case, multiple targets in range and doppler are generated, which further aggravates the range ambiguity resolver.

As shown in Figure 3.26, multiple dopplers can be easily generated by pulsing one of the amplifiers in the ECM system. The power spectrum of a CW signal transmitted by the radar is shown as a single line spectrum at its carrier frequency (Figure 3.26 a). As shown in Figure 3.26 (b), the return signal power spectrum is also a line spectrum displaced in frequency from the radar transmitted frequency by the amount of the doppler shift (a repeated signal without modulation via the ECM system will appear in the same doppler filter position as the reflected signal).

If the signal passing through the ECM system amplifiers is pulsed at doppler rates (i.e., 200 pulses/s), the radar received spectrum will appear as a line spectrum as shown in Figure 3.26 (c), with the distance between the lines equal to the PRF of the pulsing in the ECM system. In this case, the spectrum consists of spectrum lines at intervals of 200 Hz. As predicted in Fourier analysis of the signal, the rate at which the doppler power level falls off depends on the pulse length of the pulser in the ECM system; the shorter the pulse in time, the greater the number of realistic dopplers.

Extending this thought to a pulsed doppler radar, Figure 3.27 (a) depicts the power spectrum of the radar transmitted signal. As indicated, it of itself is a line spectrum with the distance in frequency between the lines exactly equal to the PRF of the radar transmitter. If the same pulsing of the ECM system signal is provided as suggested above for the CW case, the net result is another line spectrum where the ECM lines are the same as the reflected signal lines except the pulsing of the ECM signal has resulted in another set of lines about each of the original PRF lines. The difference in frequency between the lines in the new line spectrum is exactly equal to the PRF of the pulsing in the ECM as it was for the CW case.

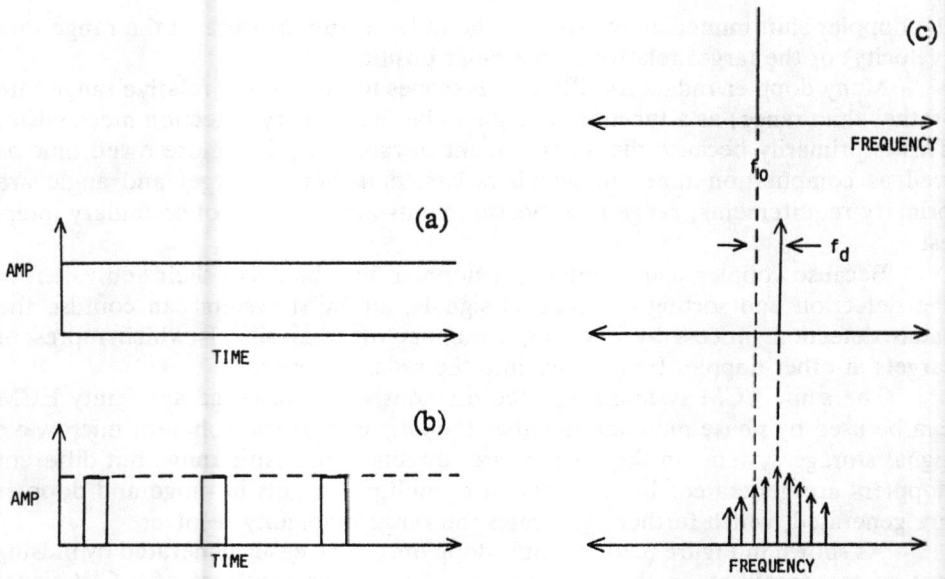

Figure 3.26 Doppler jamming of CW radar.

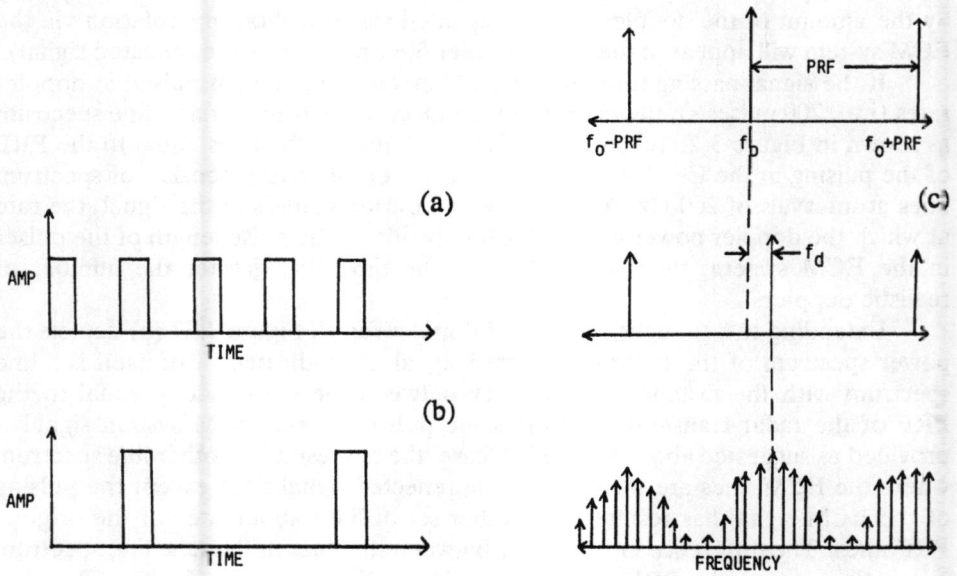

Figure 3.27 Doppler jamming of pulsed doppler radar.

The net result is a ringing of all these filters in the radar, all of which are at erroneous doppler frequencies.

3.5 ELECTRONIC COUNTER-COUNTERMEASURES (ECCM)

To preserve the integrity of target range measurement functions, radar designers have resorted to various techniques of electronic counter-countermeasures, which are designed to degrade or remove the effects of ECM. Chief among these are the techniques of carrier frequency and PRF agility. These techniques are used primarily against ECM systems which are designed to generate signals that result in noise or false targets at near-in ranges as well as at ranges greater than the true target range. Although radar antijamming techniques do not necessarily eliminate the false targets at the greater ranges, they are extremely effective for the targets at the inner ranges.

3.5.1 Carrier Frequency Agile Radars

As indicated above, if the ECM transmitted signals are triggered on one of the intercepted radar transmitted pulses and are generated into the next radar inter-pulse period, the latter signals enter the radar receiver after it is tuned to the frequency of the most recent radar transmitted signal. In the case of a radar which can change its carrier frequency on a pulse-to-pulse basis, the receiver is tuned to the most recent transmitted pulse, whereas the ECM signals entering as second-time-around signals are tuned to the prior transmitted pulse (to which the radar receiver is no longer receptive). As a result, the second-time-around signals are not detected nor displayed as targets. This is true whether the ECM signals are generated into the main beam or the sidelobes of the radar antenna (as suggested in the next chapter for angle jamming).

Figures 3.21 and 3.23 depict the effect of this ECCM technique on the PPI and B scope displays for noise or multiple target ECM systems. As indicated, all interference at the inner ranges is removed from the display, presenting only signals at greater ranges than the true target position.

A method used by ECM designers to counter this radar fix is to modulate a carrier by noise or false target pulses, the spectrum of which encompasses all frequencies within the agility range of the radar. As previously shown, this results in a dilution of effective ECM power within the detection bandwidth of the radar. This dilution is directly proportional to the ratio of radar detection bandwidth to jamming bandwidth. In many cases, because of the value of interference at the radar near-in ranges, ECM systems have been designed to produce the required ERP in spite of the demands placed on it by the dilution of power.

3.5.2 PRF Agile Radars

Another technique used by radar designers to combat the effectiveness of false targets at near-in ranges is to employ variations in the PRF of the transmitted pulses. As discussed before, STAE, as well as the signals generated by an ECM to produce inner range targets, can be made to jitter and become recognizable as false targets by varying or jittering the PRF used by the radar. This technique not only jitters unintentional STAE, but also second-time-around signals generated by an ECM system. Because second-time-around signals do not occur at the same position in the pulse interval when the radar uses PRF stagger or jitter, the ECM signals do not integrate properly on the displays or in the electronic circuits used to detect and analyze the detected signals.

The only technique available to ECM designers to counter the PRF jitter technique is against radars, for which the periodicity of the radar jitter program can be detected and measured by the ECM receiver. In this case, having determined the cyclic nature of the PRF jitter, an advanced ECM system may be able to predict the timing of the next transmitted pulse and thus ensure that the second-time-around signals in the radar are detected at the same inner range positions as required for effective inner range false target presentation. However, there is little reason for a radar to use stability or periodicity in its PRF except in the case of coherent operation, as in a doppler radar.

A more positive method available to ECM system designers to counter either of the ECCM methods discussed above is to position an ECM-carrying vehicle between the victim radar and the vehicles being protected. This is sometimes referred to as *stand-in jamming*. Although the generated noise or false targets are transmitted after interception of the radar signal by the stand-in ECM receiver, they will enter the radar receiver before the reflections from the vehicles being protected. This results in radar receiver interference occurring before the arrival of the reflections from the radar targets whether the radar PRF is jittered or not.

The use of an ECM vehicle located at a shorter range to the radar than are the vehicles being protected allows the ECM system to set on proper frequency for every pulse, even in the case of pulse-to-pulse frequency agile radars. Because the ECM vehicle must be positioned between its victim radar and the protected vehicle, this is best accomplished with an expendable unmanned vehicle. Because unmanned vehicles are generally constructed with a very low RCS, their probability of survival at near-in ranges is much better than that of a manned vehicle.

3.6 MULTIPLE RADAR CAPABILITY

This section discusses the capability of ECM systems to operate simultaneously against multiple search radars with the range jamming techniques described in this chapter. The optimum jamming system, ignoring cost, weight, volume and system complexity, is one which can isolate each of the radar signals and individually modulate each of the ECM responses with techniques which are most effective against each particular radar. This places a severe burden on a receiver associated with the ECM system to sort out the signals and to identify each as being from a specific radar as it enters the ECM system. With the advent of high-speed microminiature components both in the microwave and video frequency region, the design of modern ECM systems is approaching that required in such an optimum system.

Nevertheless, in many cases it may be possible (with little loss in effectiveness) to impose common modulations on all of the signals of any type which are passing through the ECM system. As discussed above, range jamming of search radars can be achieved by generating noise or multiple false targets which arrive at the radar before and after the reflection from the vehicle being protected by the ECM.

3.6.1 Noise

Because the purpose of noise generation against the radar's range measurement function is not necessarily to produce coherent or quasicoherent false targets, operation against multiple search radars is relatively simple. The timing of the noise transmission is relaxed because there is no attempt to present signals at predetermined positions on the radar display. Therefore, there is no special requirement for sorting or specialized modulation when this mode is used against multiple search radars in the environment. The important requirement for multiple radar noise jamming is that the carrier frequency of transmission include the carrier frequency of all intended victim radars. This can be done on a continuous basis (wideband noise jamming) or on a time-shared sequential basis (frequency-swept or frequency-hopping spot jamming).

3.6.2 Multiple False Targets

In order to present a multiple target display to a victim radar, the timing of the ECM transmission must be synchronized with the time of arrival of the signal from each radar of interest. Time modulation of the ECM transmitted signal (the transmission of signals at times delayed from the time of interception of the signal) can be applied to any number of signals as long as the timing of each transmitted signal is based on the time of arrival of each intercepted signal.

Assuming that it is possible to identify each pulse from a different radar as it is intercepted by the ECM system, a separate and distinct delay program can be applied to each of the radars. This capability allows the ECM system to present different false target displays to each of the radars if the situation warrants it. This assumption is based on an ECM receiver which is able to sort and identify each of the intercepted radar signals as they are intercepted.

Most often, specialized false target displays on each of the victim radars is not a necessity. Any false target program can be used for all of the radars as long as each program is timed on the basis of the time of arrival of each radar pulse. This mode of operation simplifies the receiver identification and sorting requirement.

3.6.3 Doppler Radar Jamming

Continuous wave or high duty cycle (on the order of 50%) signals cannot be sorted from each other as can pulsed radars on a TOA basis as suggested in the previous paragraph. Sorting of CW or high duty cycle signals can be on a carrier frequency basis if separate and distinct modulations are required on each of several such signals intercepted. The independent modulation of individual CW or high duty cycle radars can be achieved by installing relatively narrowband radio frequency filters and phase and amplitude modulators for each victim radar in the environment. If tunable filters are used, the ECM receiver must be able to identify the frequency of each radar and to tune the filters appropriately. Multiple filter banks can be used to relieve the ECM receiver tuning requirement; in this case, each filter band requires its own phase and amplitude modulators.

As described in a previous section, jamming of pulsed radars in range can be achieved by transmitting a multitude of pulses which are delayed in time (time modulated) relative to the intercepted pulse. The intent is to produce, in the radar receiving circuits, a multitude of false targets which are displayed to the operator as realistic targets at ranges other than the true target range. As indicated above, multiple targets serve to confuse the range ambiguity resolver in pulsed doppler radars. Because knowing the location of the multiple targets within the ambiguous range interval is not critical when operating against a pulsed doppler radar, the same delay program can be used on all pulsed doppler pulses passing through the ECM system with minimal loss in effectiveness against any of the radars in the environment.

Jamming in doppler (relative velocity) is achieved by one of two methods. One is to modulate the signal in phase (or frequency) as it passes through the ECM amplifier system. The other method is to modulate the radar signals in amplitude as they pass through the ECM system to generate multiple sidebands, which represent a multitude of false dopplers when processed in the radar receiver. The same modulation can be applied to all signals passing through the ECM system without loss in effectiveness.

Phase modulation for the purpose of generating false doppler frequencies can be applied without any problem, even when pulses from a conventional pulsed radar are passing through the ECM system. This is because the phase modulation is generally in the order of kHz, whereas the detection bandwidth of conventional pulsed receivers is on the order of mHz. The relatively small shift in frequency caused by the phase modulation is insignificant relative to the broad bandwidths of conventional pulsed radars. The same modulation can be applied to all signals passing through the ECM system.

3.7 SUMMARY

The objective of this chapter was to introduce the student to ECM techniques which interfere with the operation of the range measurement function of defense system radars. Unless angle jamming techniques are also included in the interference, radars may still be able to determine the angle of arrival of the reflected signal, as well as the jamming signal, if the ECM system is not collocated with the vehicle being protected. Unless the adversary employs triangulation techniques, loss of proper range measurement to the target of interest may still result.

As described in the chapter, coherent radars are more vulnerable to inner range jamming because of the inherent requirement to maintain constant frequency transmission. Furthermore, when these radars are forced to operate at high PRFs, they are not able to make range measurements on an instantaneous basis, but must complete a process required to resolve severe range ambiguities. Furthermore, if more than one signal occurs with the same ambiguous range interval at the same doppler frequency and angle, the solution generates multiple false range solutions which cannot be distinguished from the true solution.

Because noncoherent radars can be designed to change frequency on a pulse-to-pulse basis as well as use jittered PRF, these radars are difficult to counter with jamming unless the ECM system generates enough effective radiated power to spread its transmitted power over the complete range of frequency agility. When jamming in the main beam of the radar antenna, this requirement may not be severe. However, as indicated in the next chapter, the power dilution and the requirement to transmit into the sidelobes of the radar antenna may be enough to make the required ECM system impractical.

An advantage of the multiple target range confusion technique is that, with coherent microwave storage systems, true replicas of the radar signal can be generated (except at different ranges from the true target). This is subject to the limitation to outer targets when the radar employs pulse-to-pulse frequency agility. Another advantage is that the transmitted false targets for each victim radar do not necessarily occur at the same time for all of the radars. This is because there is no reason why any of the intercepted pulses should arrive at the same instant of time at the ECM receiver from multiple radars in the environment. Effectively, then, the multiple targets for one radar are interleaved with the multiple targets for another radar even though the same delay program is applied to each radar Because of the interleaving of pulse transmissions, there is little, if any, power dilution due to simultaneous transmission of multiple signals.

REFERENCE

1. Hovanessian, S.A., *Radar System Design and Analysis* (Norwood, MA: Artech House, 1984).

Chapter 4
Search Radar Angle Countermeasures

4.1 INTRODUCTION

The previous chapter discussed the techniques used by ECM systems to deny search radars the capability to determine the distance to the vehicle whose reflection was detected. Unless techniques discussed in this section are also employed, the jamming in range will occur only in the main beam of the radar antenna, which would then give the radar a measurement of the azimuth angle to the reflecting vehicle. Jamming in range only, in most cases, is not adequate to abort interception of the protected vehicle on the part of the enemy. It is more significant to deny the radar its capability to determine the angle of arrival of the echo signal, and thus the angular position of the target vehicle relative to the radar.

Because the measurement of the azimuth angle to its intended targets is most critical to a defense system, advanced radars are being designed to protect this measurement function in every way possible. In particular, radar antennas are being designed with sidelobes that are very low relative to the main beam to reduce the radar's sensitivity to signals entering the antenna outside its main beam. The value of reducing the radar antenna sidelobes will become obvious as we continue our discussion in this chapter.

4.2 SEARCH RADAR ANGLE MEASUREMENT

As indicated in Chapter 1, all radars are operated on the assumption that, if a signal enters the radar receiver and exceeds the detection threshold of the system, the signal must have been intercepted in the radar antenna main beam (i.e., the signal is received only when the antenna is pointed in the direction of the signal source). This is true whether the signal is a reflection from a target or any other source, such as an ECM system. A very strong signal which enters the antenna sidelobes may be interpreted as a main beam signal and can result in a major angle error.

Because of this vulnerability in the angle measurement system, radar designers have expended considerable effort in decreasing the radar's sensitivity to signals outside the antenna main beam. The primary effort is in the design and fabrication of antennas which produce extremely low antenna sidelobe gains relative to the main-beam gain; sidelobe levels as low as 70 dB below the main-beam gain have been realized.

Figure 4.1 shows how a search radar determines the angle of arrival of a signal detected by its receiver. As shown in Figure 4.1(a), we assume a target located at the indicated range and angle (azimuth, in this case). The actual values of these parameters are not of importance in this illustration. As shown in Figure 4.1(b), a radar is pointing at the angle indicated during the time shown. The radar is shown at the same angle in Figure 4.1(d), but, because no reflecting target is positioned at that angle, there is no power received in the antenna, either in its main beam or its sidelobes. All antennas have some degree of sensitivity at all azimuths, in the sidelobes as well as the backlobes. However, for simplicity, we assume that this antenna is completely insensitive to signals outside the pattern extremes shown. This is done without loss in generality. Therefore, at the time that the antenna is pointing in the direction shown, no signal is detected from the target, and so no power level is shown in Figure 4.1(d) at that antenna position.

No power level is indicated in Figure 4.1(d) until the antenna main-beam position is moved toward that of the target. When the antenna reaches point A, the reflection from the target enters the first antenna sidelobe and the power level depicted in Figure 4.1(d) indicates the structure of the antenna sidelobe as the antenna scans across the target position. When the antenna position reaches point B, the main beam is pointing at the target and the receiver detects maximum power at that position, which is shown in Figure 4.1(d). As seen in the figure, the power level of the signal in the receiver represents a mirror image of the antenna pattern as the antenna scans across the target position and nowhere else. If there were another target at another angle, a similar power return pattern would be shown about its angular position because of the reflections from the target at that position.

Radar detection circuits are designed to set their thresholds at power levels that can be expected from targets of predetermined cross-sectional areas and distances to the radar. This *threshold level* defines the radar receiver sensitivity and dynamic range of signal levels anticipated. If the level is set too low (too sensitive), even the signals in the radar antenna sidelobes are detected on the radar displays or in the electronic circuits. This condition is tolerable only if the operators or electronic detectors are able to determine the power amplitude variation of the detected signal as the antenna scans over the target position.

If the threshold is set too high (least sensitivity), the received power level of small targets or targets at long ranges may be so low that they are not displayed or detected in the electronic circuits. As shown in Figure 4.1(e), only during the time that the received power level is above the indicated threshold is there an

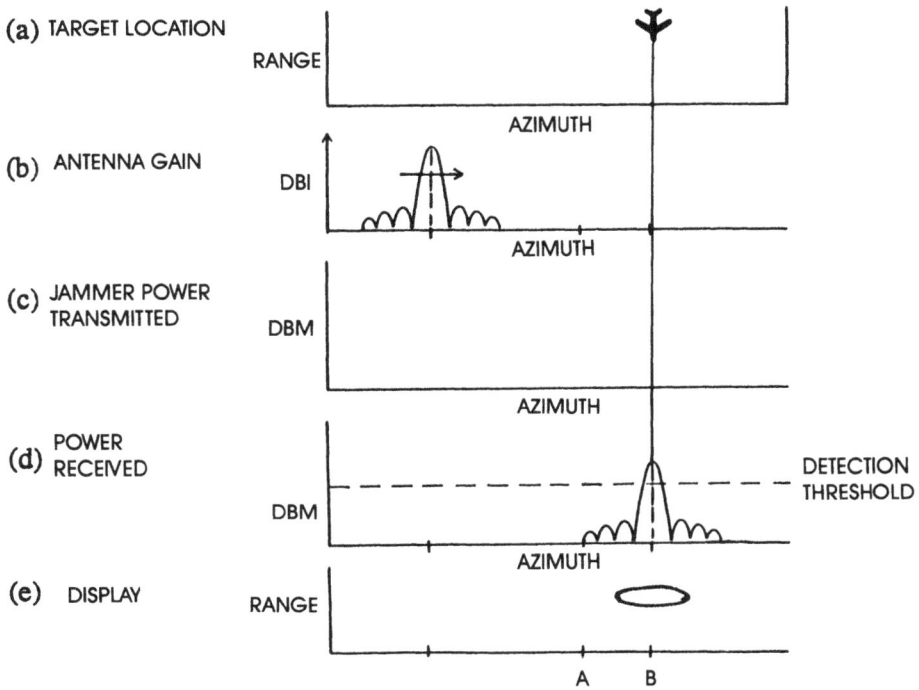

Figure 4.1 Target azimuth angle measurement.

indication on the display. The display shows a target with an elliptical shape. This is because the detected power level varies in the same manner as the antenna gain pattern with the minimum above the threshold occurring at the leading and trailing edges of the target display, and the maximum being at the center. The elliptical target pattern is especially true when the detected video is displayed on intensity-modulated screens. Radars which display targets synthetically eliminate this antenna variation during signal processing. Good operators have used this signature to identify true targets when subjected to ECM systems that attempt to confuse them with multiple false targets. This suggests an important vulnerability of synthetic display radars. Figure 4.2 shows how the raw video displays look on B and PPI scopes for the situation assumed.

We should point out that, in the situation described above, the antenna pattern used represents what is called the *two-way antenna pattern*; the signal reflected from the target depends on the antenna gain of the transmitting antenna as well as the radar receiving antenna. Assuming that the radar transmitter and

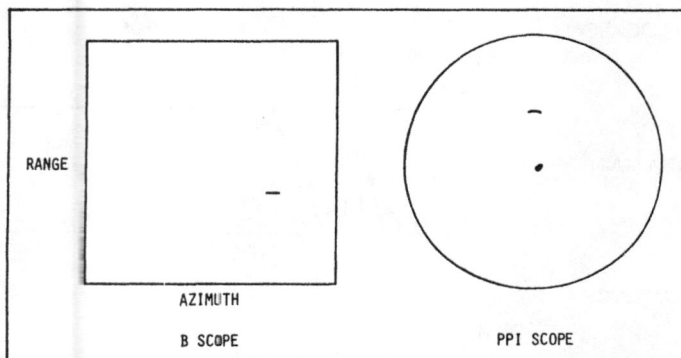

Figure 4.2 Target display on B-scope and PPI.

receiving antenna are one and the same (as is most often the case), the amplitude modulation on the reflected signal is due not only to the receiver pattern, but also the transmitting antenna pattern.

The value of deep sidelobes in the radar antenna pattern becomes evident with the two-way pattern. A reflected or repeated signal entering via a 70 dB sidelobe pattern results in received signals being as much as 140 dB less than a reflected or repeated signal located in the radar antenna's main beam. In other words, a target reflected signal in the mainbeam of a radar antenna is 10^{14} times greater than its reflection when it is received in the sidelobes of the antenna.

This result emphasizes the difficulty of ECM systems that are designed to generate signals into the radar antenna sidelobes at a level which exceeds the preset thresholds of the radar. Because these thresholds are based on signals arriving in the radar antenna main beam, signals entering the sidelobes must make up for the lack of gain at these angles. This could be as much as 70 dB greater than it would need to be in the antenna's main beam. Equation (2.17), which was developed in Chapter 2, shows the effect of this requirement with the main beam to sidelobe ratio multiplier.

The angle measurement function described is correct for all types of radars (i.e., conventional pulse, pulsed doppler, or FMCW). Although the signal processing and detection mechanism may be different in the various radars, they all operate on the same angle measurement principle: At what angle is the radar antenna pointed when detection is made? The amplitude modulation on the detected signal is used to determine the target angle, maximum amplitude occurring when the radar antenna is pointed at the target.

Although Figure 4.2 implies a continuously scanning antenna, the description applies equally well to a periodic or a nonperiodic scanning antenna system, including electronically steerable antennas which can switch from any angle to any other angle at any time and at relatively rapid rates (microseconds). A primary requirement for radar detection is that the antenna dwell long enough at each position to provide sufficient integration of the power, either on the display or in the electronic circuits.

We can assume that the antenna was switched instantaneously from position A to dwell at position B, in which case, because the antenna is not scanning, a constant power level is experienced in the radar receiver, with the power level dependent on the location of the reflecting target in the antenna pattern. Hence, the target will not assume the elliptical pattern described for the scanning antenna because the power level is relatively constant during the antenna dwell.

4.3 ANGLE MEASUREMENT ECM

4.3.1 ERP Required

To emphasize the degree of difficulty imposed on ECM systems designed to operate into radar antenna sidelobes, we will examine a typical case. For an ECM system operating into the antenna main beam, using Equation (2.12):

$$P_j G_j = P_t G_t \cdot \frac{\sigma}{4\pi r^2} \cdot \frac{J}{S}$$

where:

$$P_t = 90 \text{ dBm (1 MW)}$$
$$G_t = 35 \text{ dBi} = G_{ML}$$
$$G_{SL} = -5 \text{ dBi}$$
$$\sigma = 10 \text{ m}^2$$
$$J/S = 10 \text{ dB}$$
$$r = 100 \text{ km}$$
$$P_j G_j = 2 \text{ W, for the conditions cited.}$$

Fortunately for ECM systems (especially those that operate as constant power systems) only the one-way (receiving) antenna gain must be taken into account in the derivation of the ECM system ERP requirement. In order to produce the same radar detected power level in the antenna sidelobes, we use Equation (2.17):

$$P_j G_j = P_t G_t \cdot \frac{\sigma}{4\pi r^2} \cdot \frac{J}{S} \cdot \frac{G_{ML}}{G_{SL}}$$

where

$$P_j G_j = 20 \text{ kW}$$

As we can see, this is a direct result of the main beam-to-sidelobe gain ratio of 40 dB (35 dBi − (−5 dBi)). Nevertheless, because of the value, if not necessity, of confusing the radar as to the angle location of the vehicles being protected by the ECM system, systems are now being designed and constructed to produce the required ERP.

Constant gain ECM systems, rather than constant power systems, are used against coherent (doppler) radars. These ECM systems intercept, modulate, and amplify the radar signal before retransmitting an ECM response to the radar. As a result, a constant gain ECM system, which is designed to provide angle jamming against search radars, must be able to detect the low level intercepted signal level from the radar transmitter sidelobe antenna pattern. Furthermore, its ECM amplifier gain requirement must compensate for the two-way gain pattern loss when operating in the sidelobes of the radar. For example, whereas only 51 dB of system gain was required when operating into the radar antenna main beam (see Section 2.11), 140 dB of additional gain is required if a constant gain system is designed to operate into the sidelobes of the radar with −70 dB sidelobes—a formidable, if not impossible, requirement.

4.3.2 Constant Power ECM Transmitter

Let us assume an ECM system which has the ability to generate the power required in the sidelobes of a victim radar antenna, but provides no modulation of any kind. This capability is shown in Figure 4.3(c); in this case we see that the jammer transmitter power level is independent of the instantaneous position of the radar antenna, as we assumed. In the position shown at 0, no power is seen in the receiver because no antenna gain was assumed at this angle relative to the reflector. This remains true until the radar antenna reaches point A, when power begins to be seen in the radar receiver, as shown in Figure 4.3(d). Because of the high power level which makes up for the lack of receiving antenna gain at this point (even in the sidelobes), the power level exceeds the threshold level indicated. As we see in the figure, during all of the antenna scan pattern the power level from the ECM system exceeds the threshold, with the power level varying in accordance with the antenna pattern, as before. This results in a display which indicates target reflections that exceed the threshold, at all angles of the antenna pattern, as shown in Figure 4.3(e).

Unless the radar detection circuits or the operator can detect the variation in power due to the antenna pattern, the radar may have lost its azimuth angle measurement capability at least during the indicated antenna pattern region. However, because the intensity of the display is a function of the received power level, the intensity of the display will still vary in accordance with the antenna pattern. A good operator can detect and center on the variation of display intensity to

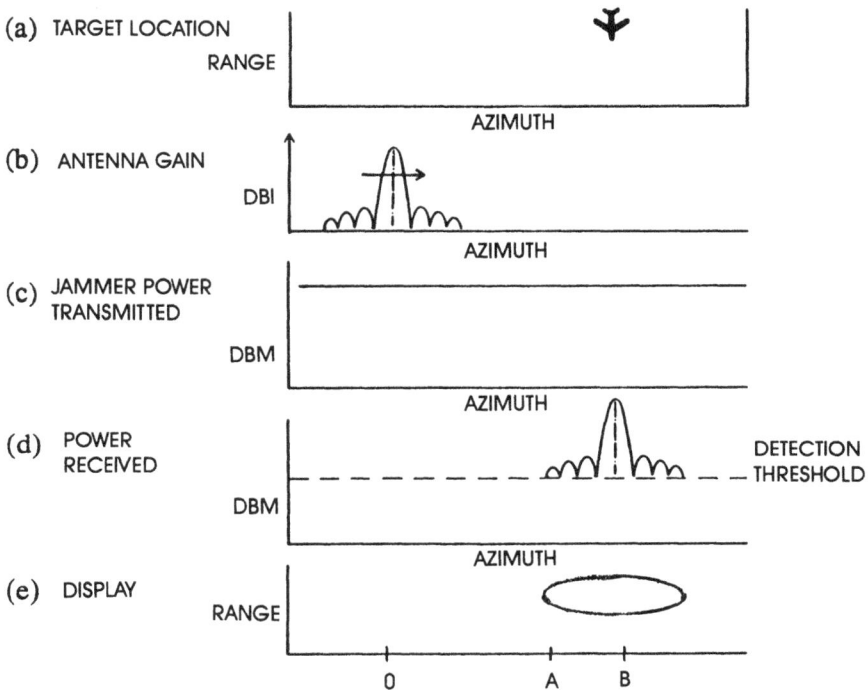

Figure 4.3 Effect of constant power jammer.

obtain a fairly good measure of the angular location of the transmitter generating the interference.

Figure 4.4 shows the display produced by a noise jamming system with the radar receive threshold setting as before; the main beam and sidelobe positions are readily detectable in both cases. Indeed, the operator may adjust the radar detection threshold so that only the main-beam power is visible on the displays. Although adjustment of sensitivity is a possible counter to high power jamming, it results in elimination of targets which may otherwise have been detected with the lower threshold.

4.3.3 Inverse Power Programming

In order to compensate for the modulation of the ECM transmitted signal by the radar receiving antenna and resultant angle measurement (as discussed in the previous section), an inverse power programming capability is usually included in

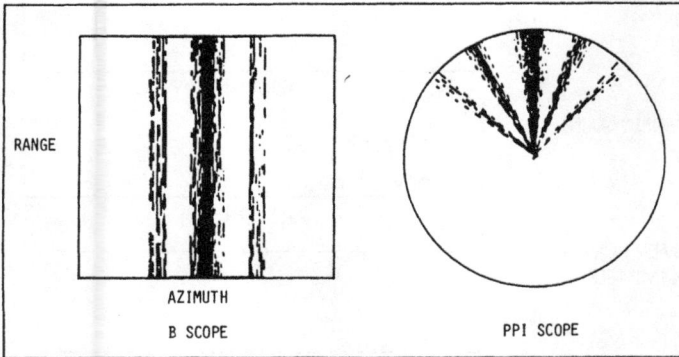

Figure 4.4 Noise jamming displays.

the ECM design. This technique requires detection of the variation in the radar signal from its transmitting antenna as it scans across the target being protected by the ECM. Because this pattern is almost always the same pattern as the radar receiving antenna, the detected pattern can be used to amplitude modulate the ECM transmitted signal in an inverse manner. As the intercepted radar signal power increases during the scanning, the ECM transmitter power is decreased in the same proportion; as the detected power decreases, the ECM transmitter power is increased accordingly.

The effect of inverse power programming is shown in Figure 4.5. As shown in Figure 4.5(c), the jammer power is constant as long as no power is received from the radar in the far sidelobes. When detection of the antenna pattern is made, the ECM system is programmed to modulate its transmitted signal in inverse proportion to that received. The net result in the radar is seen in Figure 4.5(d), which shows that the radar received power is constant for the duration of the antenna scan interval. The display in Figure 4.5(e) shows a constant power level across the total length of the antenna pattern. Although the figure indicates a constant power level in the receiver, in the real case variations will exist because of the inability of the ECM receiver to follow exactly the complex modulations in the radar antenna pattern.

Figure 4.6 is another representation of this concept. In Figure 4.6(a), the radar signal (as received by the ECM system) varies in accordance with the antenna pattern of the radar transmitter; in Figure 4.6(b), the amplitude of the ECM transmitted signal is controlled in an inverse fashion to that of the received signal. The net result (as detected at the radar receiver) is shown in Figure 4.6(c) as a series of pulses with amplitude that is constant and independent of the radar antenna pattern. As can be seen in the figure, the effect of a noise jammer with

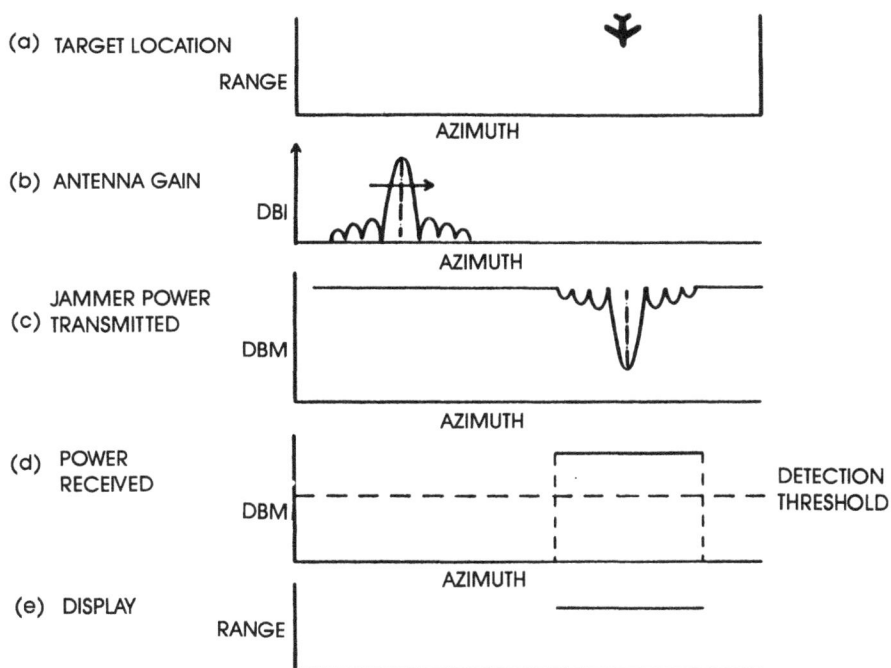

Figure 4.5 Effect of inverse power programming.

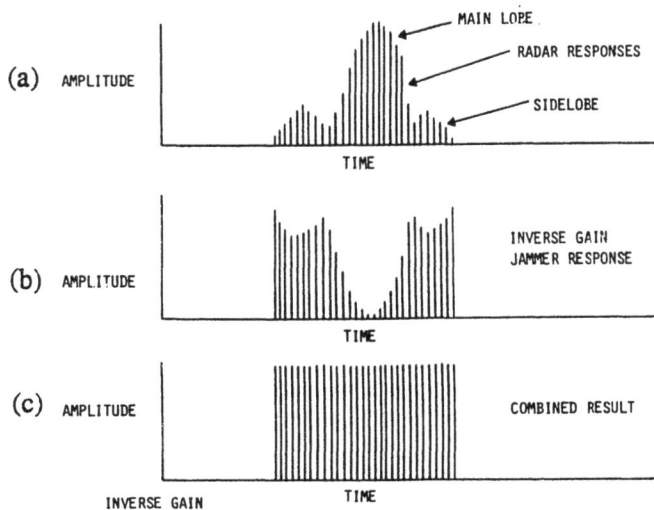

Figure 4.6 Inverse power program.

inverse power programming is to produce a noise strobe as wide as the antenna pattern (including the near sidelobes) yet independent of the radar antenna variations in that region.

Although the PPI and B-scope displays (Figure 4.7) show the effects of noise jammers, the same concepts apply to multiple range target jamming systems. Multiple targets in range can be generated by delaying the intercepted signal (in a programmed fashion) to yield targets at greater ranges than the true target; if these are delayed into the next radar interpulse period, some of these will appear as targets nearer to the radar than the true target, assuming no radar antijamming techniques such as carrier frequency or PRF agility are used.

Multiple targets in range can be used equally well when transmitting into the radar sidelobes. This mode of operation will produce multiple targets in range as well as at different angles. Figure 4.8 shows the displays when an ECM system is designed to generate targets in range and in azimuth. In order to remove the power modulation due to the scanning radar antenna pattern, inverse power programming must also be applied to each of the false targets.

Because the size in angle (width) of each target (as seen on the operator's display) is dependent on the number of pulses transmitted at each delay in the ECM system, the modulation program can be adjusted to provide different sizes of targets at different ranges and angles. This results in a wide range of different sizes of targets which makes differentiation from the true target more difficult for an operator. The amplitude of the pulses at each of the range and angle positions can be programmed to produce elliptical targets of varying degrees of amplitude and size, if necessary.

Figure 4.9(c) shows a type of amplitude modulation which can be superimposed at one false target range position. With current technology, producing a wide variety of targets at different ranges and angles is possible with different amplitudes and sizes. The ability of an operator to identify the true target in this type of presentation is highly questionable.

4.3.4 Doppler Radars

As we have indicated, to inject signals into the sidelobes of a radar antenna with sufficient magnitude to exceed the detection threshold in the radar, the ECM transmitter power level must be increased to compensate for the lack of antenna gain in the sidelobes. This leads to an ECM transmitter requirement for exorbitant amounts of power and, in some cases, impractical power levels. This is especially so when operating into a conventional pulse radar, which cannot and does not attempt to compete with ground clutter or natural interfering signals.

The function of doppler radars, on the other hand, is to be able to extract signals when competing with very high clutter returns. This is achieved by differentiating between the clutter returns and targets moving relative to the clutter

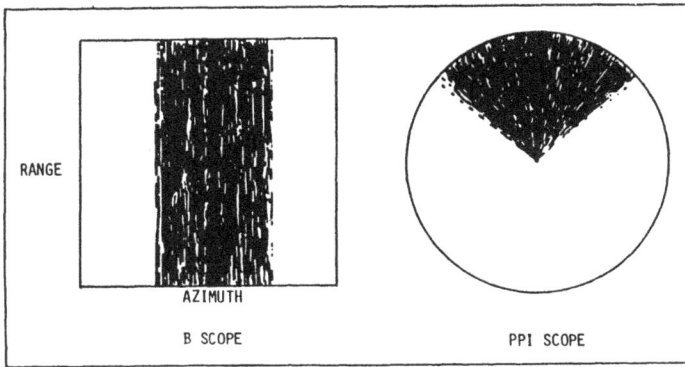

Figure 4.7 Noise inverse power jamming display.

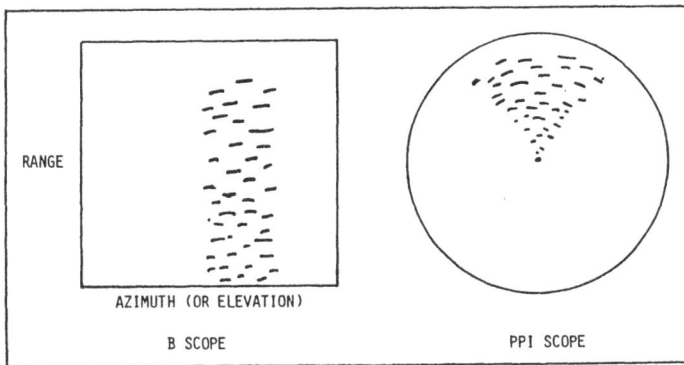

Figure 4.8 False target inverse power jamming display.

based on very small differences in the shift of the radar carrier frequency due to the doppler effect of moving targets. As a result, the doppler radar cannot present "raw video" to the operator because the extraction process is performed electronically and most often with digital signal processing, during which important raw video characteristics are lost.

Furthermore, because of the large size of the clutter return relative to expected signal returns, it is imperative that the doppler radar maintain a very high dynamic range of operation in the amplitude of the signal returns. The range of signal returns may be as large as 30 dB, whereas the size of the clutter can be as much as 60 dB above the largest signal return. This places a requirement for a very high instantaneous dynamic range of operation in coherent radar receivers.

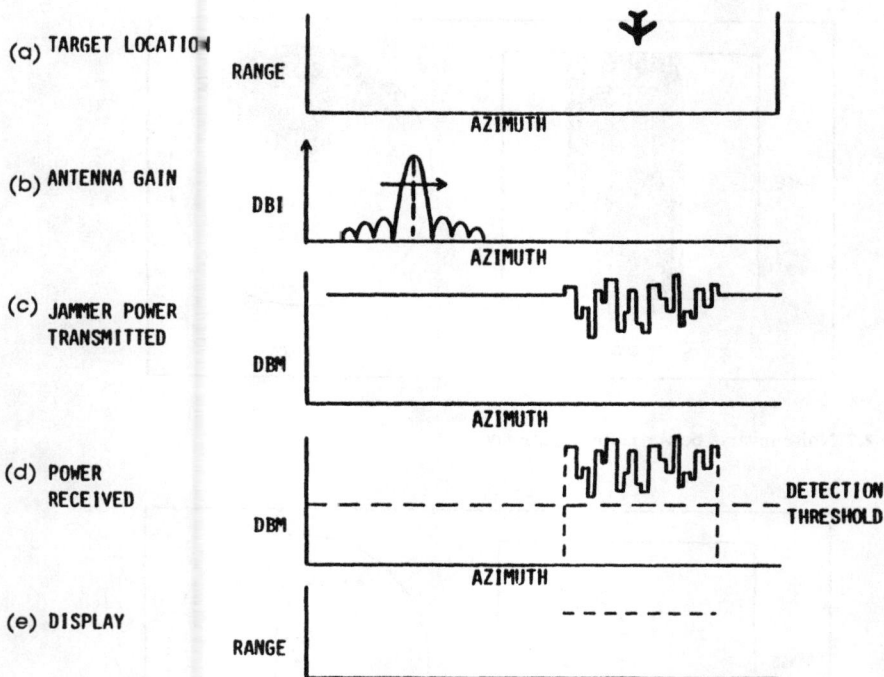

Figure 4.9 Effect of programmed false targets.

Because of these requirements, the doppler radar is almost always designed to detect and analyze all signals which exceed the preset detection threshold of the radar. This argument relaxes the ERP requirements of an ECM system designed to inject signals into the antenna sidelobes of a doppler radar system.

For the conventional pulse radar, the ECM ERP produces a signal in the sidelobes of the radar antenna with intensity equal to a signal detected by the radar in the main lobe of the antenna. The ERP requirement is increased directly as the main beam-to-sidelobe gain ratio. However, for doppler radar, an interfering signal injected into the sidelobes of the radar antenna need not be of the same level as a signal detected in the mainbeam. We need only transmit a signal of sufficient power that, when it reaches the radar detection circuits, it exceeds the detection threshold of the radar receiver. In most cases for a doppler radar, this is the most sensitive level of the receiver. Using the equation:

$$P_j G_j = S_{\min} \cdot \frac{(4\pi)^2 r^2}{G_{SL}\lambda^2}$$

$$(4.1)$$

where S_{min} is the minimum sensitivity of the radar receiver. For a radar with minimum sensitivity of -100 dBm, which is typical of doppler radars, we obtain:

$$P_j G_j = 50 \text{ W, for } r = 100 \text{ km and } \lambda = 3 \text{ cm}$$

This is a much more realizable value of ERP for an ECM system designed to generate interference into the sidelobes of the radar receiving antenna.

An interesting result of this concept is shown in Equation (4.1), which indicates that the ERP required for sidelobe jamming into doppler radar sidelobes decreases as the square of the range. Assuming a constant ERP, the J/S increases by 6 dB as the range decreases 3 dB. This is the direct opposite of the conventional pulse equation, but it is more favorable for coherent radar jamming.

4.4 ELECTRONIC COUNTER-COUNTERMEASURES

4.4.1 Introduction

As we have discussed in Section 1.4.2, a serious vulnerability of all search radars is that the radar angle measurement is based on the fact that, if a signal is received and detected by the radar, the source of the signal (reflected or generated by an ECM system) must be in the main beam of the radar antenna. Therefore, by noting (mechanically or electronically) the position of the antenna and its pointing angle at the time of detection, the radar determines the angular position of the source relative to the radar. However, as we have discussed, if a signal has a power level high enough to exceed the radar detection threshold, even when it is injected outside the main beam of the radar antenna, the radar concludes that the source of that signal was in the mainbeam and thus at the angular position defined by the mainbeam. This is the radar vulnerability that is exploited by the angle jamming techniques discussed in Section 4.3.

4.4.1.1 Ultralow Antenna Sidelobes

To reduce or remove this vulnerability in radar angle measurement, radar antenna designers have developed antennas with ultralow sidelobes (as much as 70 dB). As shown in the previous section, extremely low sidelobes in victim radar antennas place a formidable, if not impossible, requirement on the design of ECM transmitter systems.

4.4.1.2 Blanking

To counter sidelobe jamming techniques, radar designers have resorted to techniques of *sidelobe blanking* and *sidelobe cancelling*. In these techniques, an auxiliary antenna is used, as shown in Figure 4.10. The A antenna is the main antenna (as previously defined) with the mainbeam and sidelobe structure designed to yield the target displays required; the B antenna is the auxiliary antenna needed to provide the blanking and cancelling.

As shown in the figure, the gain of a signal in the mainbeam of the A antenna, is greater than the gain in the B antenna. For a signal anywhere in the sidelobes of the main antenna, the gain of the B antenna is greater than that of the A antenna. As shown in Figure 4.10, each of these antennas is connected to its own receiver. The amplitude levels are then compared at the output of the receiver. If the signal in the A receiver is greater than in the B receiver, the radar correctly concludes that the signal entered the receiver with the antenna pointed at the target; the signal is then gated through to the signal analysis circuits. However, signals the amplitude of which is greater in the B receiver than in the A receiver are prevented from entering the signal analysis circuits. This process is continued on a pulse-to-pulse basis for a conventional pulse radar.

Therefore, when this radar antijamming technique is used, pulses generated into the radar antenna sidelobes, as suggested as an angle ECM technique, result in signals with greater amplitude in the B receiver than in the A receiver, which are prevented from entering the signal analysis circuits and the radar display. In this case, the only degradation to main beam target detection would be when one of the sidelobe-generated pulses arrives at the radar at the same instant as a true signal in the main beam. The probability of this occurring is very small (dependent upon the radar transmitter duty cycle). It is difficult, if not impossible, for an ECM system to force the pulse coincident condition because of the two different locations of the main beam target and the sidelobe generating source relative to the victim radar.

As shown in Figure 4.10, the blanking in the radar is performed on a pulse-to-pulse basis, on the time of arrival of the pulses at the two receivers. In doppler radars, the blanking of signals is performed on a doppler frequency basis rather than time of arrival. This is because doppler filtering is the primary sorting mechanism in doppler radars. As shown in Figure 4.11, if the signal in the doppler filter at the time of detection is greater in receiver A than that in the same doppler filter in receiver B, it is gated through the doppler gate to the analysis circuits; otherwise, it is not allowed to pass through to the detection and processing circuits because the signal is determined to have arrived via the radar antenna sidelobes.

The same argument holds for the doppler case as to the probability of blanking out signals which arrive via the main beam at the same time as the interfering doppler signal. However, the probability of simultaneous detection is much more

SIDELOBE BLANKING - CONVENTIONAL PULSE.

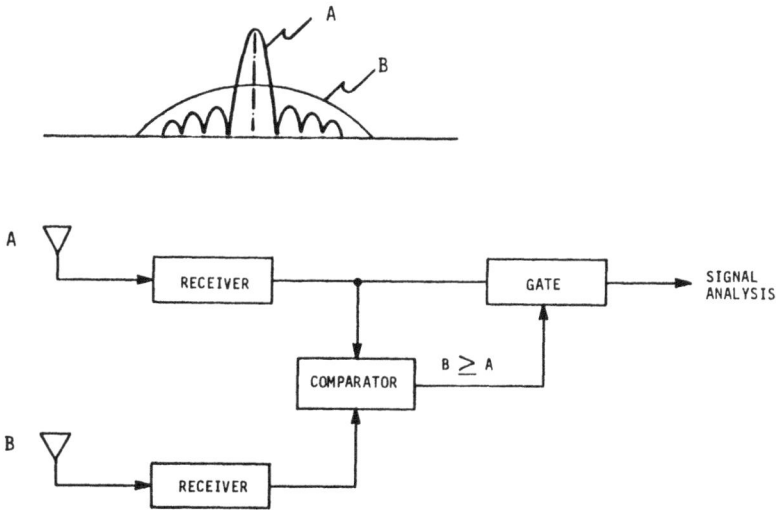

Figure 4.10 Sidelobe blanking—conventional pulse.

SIDELOBE BLANKING - PULSE DOPPLER

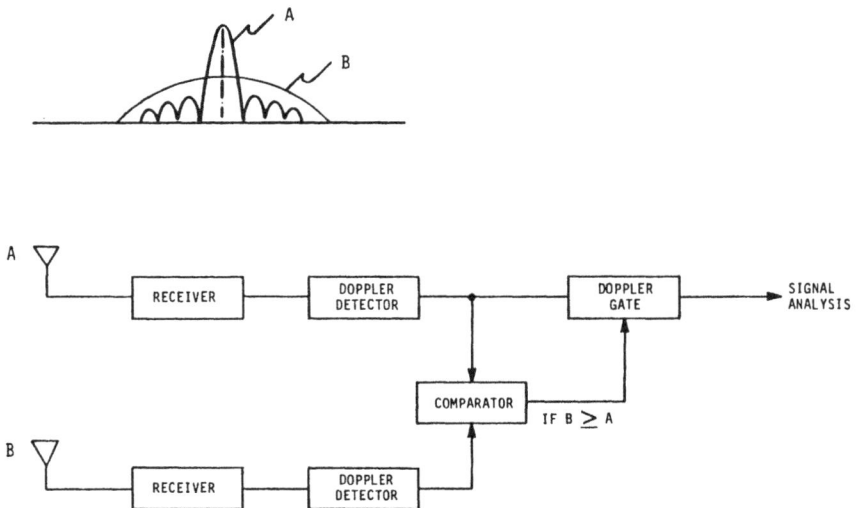

Figure 4.11 Sidelobe blanking—doppler radar.

favorable in this case, because signals at multiple doppler frequencies are easily generated, as discussed in Section 3.4.

4.4.1.3 Sidelobe Cancelling

Because the ECCM blanking technique does not apply to a continuous wave or noise interference, the cancelling technique shown in Figure 4.12 was developed. In this case, the interfering signal is detected in both the A and B receivers, and if the power level is greater in B (indicating that it arrived via the radar antenna sidelobes), a cancelling process is initiated. In the cancelling process, the amplitude and phase of the interfering signal are adjusted in a closed loop via the nulling feedback circuit so that the interfering signal is minimized in the main receiver channel.

The intent of this technique is to reduce the interfering signal level in the main channel and thus improve the signal-to-noise ratio (S/N) for the signal arriving via the main beam in the A receiver. Cancelling ratios as high as 30 dB have been achieved with this method. A serious deficiency of this technique is that a separate canceller is required for each interfering source in the radar antenna sidelobes. A total of five cancellers was proved to be the practical limit for effective cancelling with this technique.

4.4.1.4 Cross-Polarization Techniques

In order to counter the blanking and cancelling ECCM techniques, ECM designers have suggested techniques of cross- or near cross-polarization of the signals transmitted into the radar antenna sidelobes. *Polarization* is defined as the orientation of the electric vector of a radiated signal; orientations include vertical, horizontal, fixed slant, circular or elliptical. As shown in Figure 4.13, for a typical search radar, a signal injected into the antenna at a polarization other than the antenna's design polarization can disrupt the antenna patterns in a way that destroys the criteria required for effective comparison of signals in the main and auxiliary antennas. This is because the two radar antennas (main and auxiliary) are two physically different antennas, even though one may use elements of the other to execute the technique. In most cases, the two radar antennas are sufficiently different, and so produce different polarization characteristics, which make them vulnerable to cross-polarization techniques.

In Figure 4.13, the solid line patterns represent the correct or properly polarized case for a typical radar antenna; the dotted lines represent the cross-polarized case (the ECM transmitter antenna polarization is orthogonal to the radar receiving antenna polarization). For the co-polarized case, the relationship of the two patterns is correct for proper blanking or cancelling operation; for the cross-polarized case, the patterns show a reversal of the roles of the two antennas.

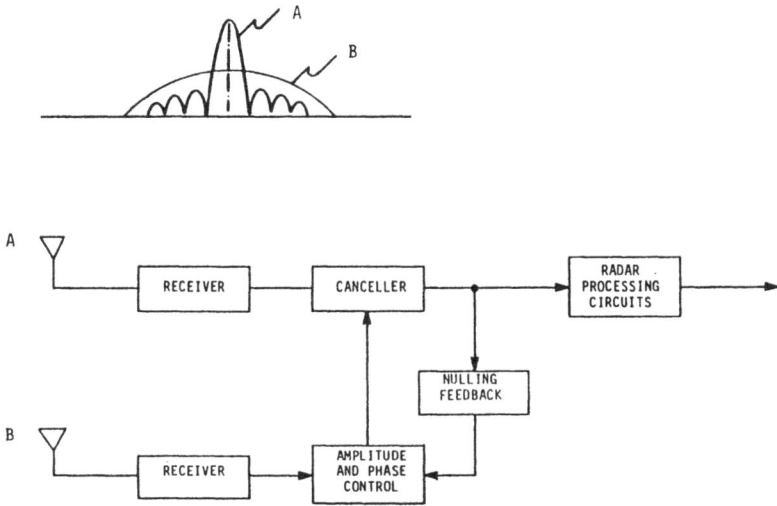

Figure 4.12 Sidelobe cancelling.

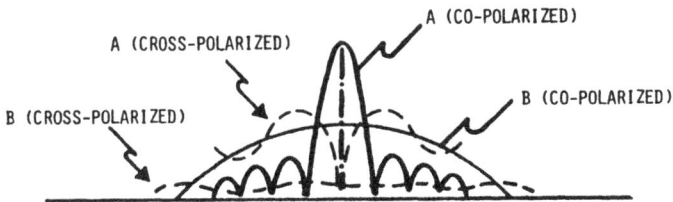

Figure 4.13 Antenna polarization patterns.

For a cross-polarized signal in the sidelobes, the main antenna provides more gain than that in the auxiliary antenna. Therefore, signals arriving via the antenna sidelobes and with a polarization orthogonal (90°) to the design polarization of the antenna will pass through to the analysis circuits.

We should point out that the cross-polarization technique is effective if the ECM transmitted signal is orthogonal or nearly orthogonal to the radar receiving antenna. ECM receivers detect and measure the polarization of the radar transmitting antenna, which may or may not be the polarization of the radar receiving antenna.

4.4.2 Electronically Steerable Antennas

To modulate the ECM transmitted signal properly in amplitude for the purpose of angle jamming of search radars, the ECM system must be able to determine the scan characteristics of the radar antenna, in particular, its pattern and scan rate. This is especially true when it is required to synchronize the false target positions on a scan-to-scan basis. This is accomplished by detection of the amplitude variation of the intercepted radar signal as the antenna is scanned across the ECM vehicle position.

Mechanically scanned antennas are generally constrained to scan in a periodic fashion, which may be detected at the ECM receiver. However, electronically scanned antennas can switch at very rapid rates and into random angular positions. Under these conditions, it is not possible to reliably provide amplitude modulated angle jamming techniques such as inverse power programming (discussed in section 4.3.3).

4.5 ANGLE ECM MULTIPLE RADAR COMPATIBILITY

To provide the angle jamming capability described in the foregoing paragraphs against multiple radars simultaneously, the signals must be modulated separately so that the modulations are uniquely applied to each victim radar signal. This is because each radar will produce different measurable antenna characteristics, depending on the design of the antenna and location of the radar relative to the ECM vehicle.

4.5.1 Inverse Power Programming

The *inverse power programming* technique is based on the ability of the ECM receiver to detect the pattern of the scanning antenna. Because we can expect that the patterns (as well as the occurrence of the antenna main beam in time) are unique for each radar, separate and different amplitude modulations of the ECM transmitted signal are required. It is not possible for an ECM system, in its angle jamming mode, to use an amplitude modulation which is effective against more than one radar at a time. For an optimum response in the case of multiple victim radars, it is necessary for the ECM system to separate the signals from each other, measure their individual characteristics, and to sequentially (on a time-shared basis), modulate the ECM transmitted signal. The time-sharing can be done on a pulse-to-pulse or angle-to-angle basis.

4.5.2 Cross-polarization

Cross-polarization, the technique producing a phase front distortion for the purpose of angle jamming search radars, dictates an ability to measure the polarization characteristic of each victim radar. Therefore, the ECM receiver must measure the polarization of each radar and the ECM transmitter polarization must be appropriate for each radar in the environment. This can be accomplished on a pulse-to-pulse or time-shared basis.

4.6 RANGE, DOPPLER, AND ANGLE ECM COMPATIBILITY

Although the previous sections individually described the techniques of range jamming, doppler jamming, and angle jamming, we do not imply that each of these techniques must be used individually. For most effective interference with radar operation, we recommend that a combination of these ECM techniques be employed simultaneously. Furthermore, because of the large number and large variety of types of radars anticipated in the environment (many of these in the same engagement), the ECM system should be able to operate simultaneously against as many radars as possible with a combination of techniques.

Compatibility between the various ECM techniques is achievable when the types of modulation required in each are compatible. Range jamming depends on time modulation of the transmitted signal; doppler jamming can be provided with phase (frequency) modulation or amplitude modulation; other than the cross-polarization technique, angle jamming relies on amplitude modulation of the ECM transmitted signal. All three modulations (time, phase, amplitude) can be simultaneously imposed on the ECM signal with the type of multimode system shown in Figure 4.14 without compromising the effectiveness of any of the techniques.

With the advent of high speed logic and switching circuits, a multimode system (as shown in Figure 4.14) can independently and uniquely impose all modulations (including antenna modulation techniques such as cross-polarization), on a pulse-to-pulse basis. Each pulse transmitted can be modulated uniquely in amplitude, frequency, and time, as well as polarization, using techniques which are most appropriate for the intended victim radar. This capability depends on the ability of the ECM receiver to sort out and identify each of the radar signals intercepted, and to predict the *time of arrival* (TOA) of each radar pulse at the receiver.

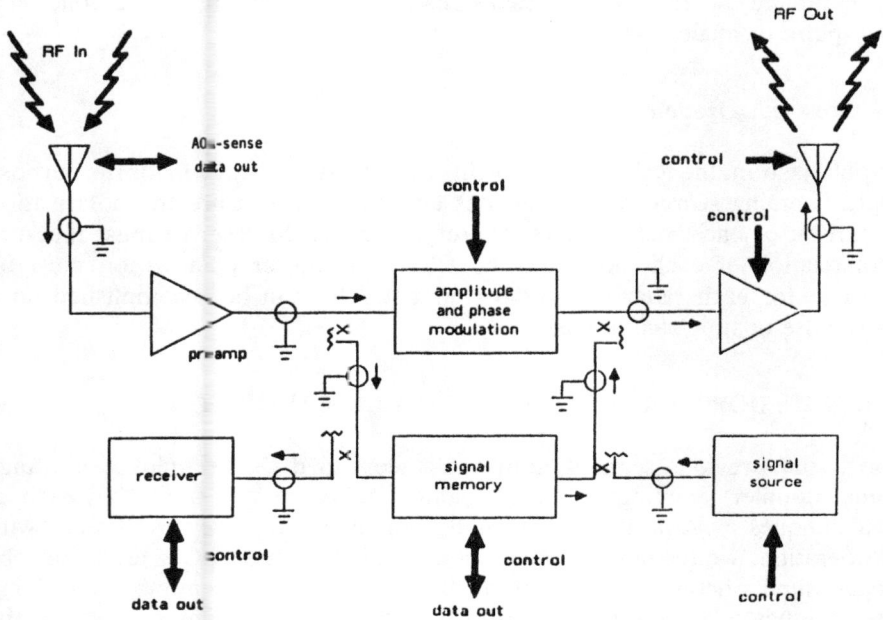

Figure 4.14 Typical ECM system architecture.

Extensive effort is currently being made to provide a capability which would allow simultaneous and unique modulation for each victim radar. The alternative is to jam only the radars which are of the most immediate threat to the mission of the protected vehicle.

As described in Section 3.3.2.1, jamming of pulse radars in range can be achieved by transmitting a multitude of pulses which are delayed in time (time modulated) relative to the intercepted pulse. The intent is to produce a multitude of false targets in the radar receiving circuits, which are displayed to the operator as realistic targets at ranges other than the true target range.

Jamming in doppler (relative velocity) is achieved by one of two methods. One is phase modulation (frequency modulation) of the signal as it passes through the ECM amplifier system; the other method is amplitude modulation of the radar signals as they pass through the ECM system. Both types of modulation generate multiple sidebands, which represent a multitude of false dopplers as processed in the radar receiver.

Phase modulation for the purpose of generating false doppler frequencies can be used at all times without any problem, even when pulses from a conventional noncoherent radar are passing through the ECM system. This is because the frequency modulation is generally on the order of kilohertz, whereas the detection bandwidth of convention pulse radar receivers is on the order of megahertz. The relatively small shift in frequency caused by the phase modulation is insignificant relative to the broad frequency bandwidths of conventional pulse radars.

Continuous wave or high duty cycle (on the order of 50%) signals cannot be sorted out from each other as can pulse radars on a TOA basis, as suggested in the previous section. Sorting of CW or high duty cycle signals must be based on a carrier frequency basis if separate and distinct modulations are required on each of several such signals intercepted in the environment.

The independent modulation of individual CW or high duty cycle radars can be achieved by installing relatively narrow-band radio frequency filters, and amplitude and frequency (phase) modulators for each victim radar in the environment. If tunable filters are used for separation of the incoming signals, the ECM receiver must be capable of identifying the frequency of each radar and appropriately tuning the filters. Multiple filter banks can be used to relax the ECM receiver tuning requirement; in this case, each filter band requires its own phase and amplitude modulators.

Nevertheless, amplitude modulation of the signal passing through the ECM system used to generate multiple false doppler frequencies can seriously affect the amplitude modulation, simultaneously producing angle jamming of the victim radar. This is because the amplitude modulation used for doppler generation is in the same bandwidth as that which is used for angle jamming. This incompatibility is aggravated by the fact that the amplitude modulations generally used in ECM systems for either type of jamming are on-off types. If this incompatibility were not recognized, the simultaneous amplitude modulation could possibly result in the ECM transmitter radiating at such a low rate that neither of the modes of jamming would be effective.

4.7 SUMMARY

In this chapter we discussed various techniques of ECM that exploit the inherent vulnerability of radar systems in their measurement of the AOA of detected signals. Because these techniques are based on the ECM system transmitting into the radar antenna when it is pointed away from the ECM system, extremely large amounts of ERP are required in order to overcome the main beam-to-sidelobe ratio of the victim radar antenna.

Because the angle parameter of the radar target's location is so critical to the defense problem, ECM systems are being designed and produced despite this high ERP requirement. Radar designers, in turn, are just as determined to preserve this capability, with the development of ultralow sidelobe antennas, and antijamming techniques such as sidelobe blanking and cancelling.

Because of the high ERP requirement in ECM systems to effect jamming in the sidelobes of radar antennas, special vehicles, surface or airborne, are being used in an escort, stand-off or stand-in role for the sole purpose of generating the jamming signals required for protection of the penetrating vehicles. The dedicated vehicles are equipped with very high power transmitters and electronically steerable antennas, as well as sophisticated receivers for effectively managing the large but limited power available.

Use of the support vehicles to produce the required jamming into the radar antenna sidelobes relieves the penetrating vehicles from that responsibility. The penetrating vehicles can then concentrate on the radars which are terminal threats to their survivability. These threats include radars which are used to direct interceptors (manned or unmanned) to intercept the penetrating vehicles. In the next two chapters, we discuss the ECM techniques used to interfere with the proper operation of terminal threat radars.

Chapter 5
Tracking Radar Range Countermeasures

5.1 INTRODUCTION

A tracking radar, as defined in this book, is one that uses electronic techniques to determine the angular location of targets of interest to within a fraction of the radar's antenna beamwidth. In most cases, the tracking radar also determines accurate location of targets in range by using electronic techniques, although this parameter is of secondary interest in some applications. The tracking radar uses each measurement of target position to plot target movement to determine the intentions of the target and to determine the parameters required for proper guidance of interceptors.

Tracking radars are generally designed to dwell long enough on their targets to yield a signal data rate that results in target location measurements sufficiently accurate to guide manned or unmanned vehicles to successful interception of the targets. In some cases, such as for control of gunfire, accurate measurements are required in both the range and angle parameters. In other cases (such as command guidance of missiles or illuminators for semiactive homing guidance), measurements in angle alone may be adequate, in which case, location accuracies in angle need only be good enough to maintain the target within the beamwidth of the radar antenna. Because the tracking radar which is required to provide accurate measurements in both the range and angle parameters is the more common case, this chapter and the next concentrate primarily on that type of radar. This chapter discusses techniques of radar range tracking and ECM techniques developed to interfere with that function. The next chapter discusses techniques of radar angle tracking and ECM used to interfere with that function.

Tracking radars are extremely important in defense systems that rely on them to provide data for effective launch and guidance of the intercepting vehicles toward targets of interest. To perform this function effectively, the radar is required to measure accurately the positions of targets designated by other sensors, such as early warning and acquisition radars. Even though the interceptor may possess its own terminal guidance to the target, the limited field of view of the terminal guidance antenna still demands accurate cueing of the interceptor guidance system by the tracking radar.

To provide the required accuracy in the measurement of target position, these radars are designed with very narrow transmitter pulsewidths and narrow antenna beamwidths to improve the target resolution in range and angle, respectively. Whereas search radars employ pulsewidths on the order of microseconds (which is more than adequate for their responsibility), the tracking radars use pulsewidths that measure in fractions of microseconds. Although advanced radars using pulse compression can transmit and receive pulses of extended widths (10 μs or more), processing in the radar receiver resolves targets and measures range to within a fraction of a microsecond.

The antenna beamwidths associated with search radars are on the order of several degrees, whereas tracking radars use antennas with beamwidths of two degrees or less. Because the primary responsibility of search radars is early detection and warning (and not necessarily accurate target position measurements), wider antenna beamwidths are tolerable in these radars.

In addition to the narrow pulsewidths and narrow antenna beamwidths used in radar design to improve the resolution and accuracy of target position measurements, tracking radars must operate with a very high data rate of reflected signals from any one target. As statistics have shown, the more samples of data that are used, the more accurate and reliable is the data measurement. The high data rate requirement dictates that the tracking radar antenna dwell on the target for a much longer period than can be expected or tolerated in a search radar, which has a very large volume of space to search. As a result, the space volume of responsibility is more limited with tracking radars, and so too is the number of targets and their position parameters that can be measured at any one time.

With a single target responsibility, the tracking radar usually spot lights the target by continuously pointing its antenna main beam in the direction of the target. In this mode of operation, the radar continuously measures the position and motion of its target so as to be able to predict its position in range and angle on each subsequent return. This is required to keep a moving target within the radar's narrow beamwidth and range gate. The data is also used to determine the intentions of the target and the optimum response to react to its presence.

When multiple-target tracking is required, the radar is forced to time-share the antenna main-beam positioning as well as the radar processing. The extent of this sharing is controlled to prevent compromising the measurement accuracies to the point that ineffective operation of the defense system results. Simultaneous tracking of several targets in angle is discussed in the next chapter.

The discussion here is based on the assumption that another sensor, such as an early warning radar, an acquisition radar, or the search function associated with a tracking radar, has specified the location of the target (with fairly rough estimates of the target's azimuth and elevation angles), and the range from radar to target. Because the tracking radar can search the complete range of interest very rapidly (within microseconds), target range designation may or may not be made by the cueing sensor. However, there is some decrease in radar performance when range designation is unavailable or inaccurate, because the rapid scan of the range gate over a large interval gives less opportunity for signal integration. Modern systems correct for this by using a multiplicity of contiguous range gates, but the result is a higher false alarm rate, and may lead to acquisition of false targets or the wrong target.

5.2 TRACKING RADAR RANGE MEASUREMENT

5.2.1 Pulse Range Tracking

Whereas search radars need only measure the range to a target to within a pulsewidth, range tracking radars are designed to measure the range to the targets to within a fraction of a pulsewidth to satisfy the requirement for accurate guidance of manned or unmanned interceptors to their intended targets. Such tracking radars are designed to measure and to maintain track on a specific part of the received target pusle, rather than on just any part of the pulse.

In pulse tracking radars, the point of the pulse to be tracked and measured is generally the centroid of the energy contained within the pulse received. This is accomplished as shown in Figure 5.1. The radar uses two contiguous range gates, the total width of the two being slightly greater than the radar transmitted pulsewidth. The first of the two gates is called the *early gate* and the second, the *late gate*.

As shown in Figure 5.1 (b), the total pulse is channeled into each of the gate circuits in time synchronism. The comparator then compares the amount of energy in the early gate with that in the late gate. If the gates are not centered on the pulse energy centroid, an error signal is produced. The amplitude and polarity of the error signal define how far and in which direction the gates must move to equalize the energy in the two gates.

With a target that has no range change with respect to the radar (*zero relative velocity*), the error signal tends toward a null position eventually, depending on the characteristics of the tracking loop. This drives the early and late gates to a position where the energy levels are equal in both, which is the point of the pulse used to measure the range to the target. Generally, these tracking loops operate with some error, depending on the target range rate relative to the radar. Such tracking loops are also designed to operate with memory, which allows them to

AUTOMATIC RANGE TRACKING

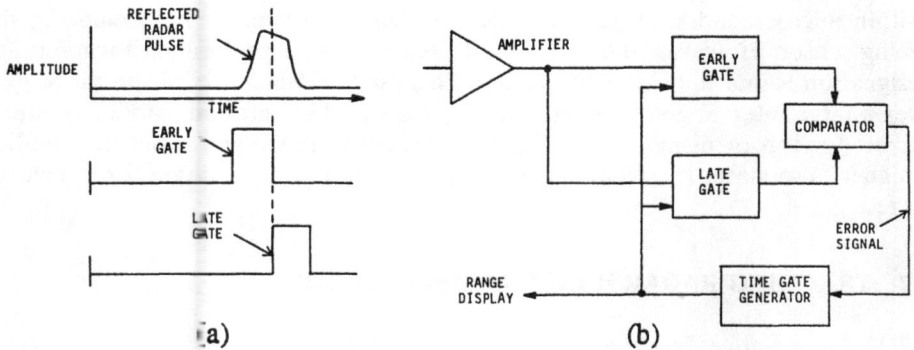

Figure 5.1 Automatic range tracker.

coast at the last measured range rate during periods of lost data; such a loss of data can be due to intentional or unintentional interference, including fading of signals.

The position of the time gate generator is initially set by the operator or the electronic circuits based on data from early warning radars, acquisition radars, or other sensors. If a designation is not made as to the range position of the target, a range sweep can be initiated, which encompasses a search of the complete suspected range of the target, and even the complete radar range of interest, if necessary. Because a search of the complete range at any one angle can be performed in milliseconds, little degradation of the performance of the radar operation in its acquisition process results if designation of target position is not made.

During the range gate search, the range tracking loop is generally designed to accept a transient because the motion of the tracking gate in its range search mode is detected as additional motion on the target, even if the target is stationary relative to the radar. Radar designers include a broad acceptance bandwidth in the acquisition process and switch to a bandwidth (on the order of hertz) that is compatible with the actual dynamics expected in the environment during the tracking process. The more narrow is the bandwidth, the more accurate is the tracking and thus the measurement of the target position.

Furthermore, the narrow bandwidth of the range tracking loop protects the system from interference which comes with sudden changes of relative velocity, as experienced with chaff drops. For the same reason, false targets generated to deceive the range tracking loop must exhibit acceleration levels (changes in relative velocity) which are within the acceptable limits of the tracking loop. We emphasize that added motion on an ECM signal, either in relative velocity or acceleration, is algebraically added to that which already exists on the vehicle where the ECM system is located.

In addition to the early and late gates, the tracking gate generator provides a *tracking gate,* the width of which is slightly larger than the total width of the two contiguous gates, and which is in synchronism with, and overlaps, the two gates in time. This tracking gate is used to isolate the target of interest from all other targets which may be located at the same angle but different ranges. The use of the tracking gate also eliminates interference (intentional or unintentional), located at ranges before and after the *time of arrival* (TOA) of the signal of interest, from the detection circuits of the radar.

The fact that the tracking gate also determines which signals enter the angle tracking gate of the radar is extremely important to an ECM system. No signals other than those contained within the radar range tracking gate are processed in the angle tracking loop. The radar tracking gate is also used to isolate the desired target signal for the *automatic gain control* (AGC) loop, which adjusts the gain of the radar's amplifliers to a level commensurate with the power level of the tracked signal. Therefore, if an ECM system is designed to interfere with the range tracking, angle tracking, or AGC loops, it must be capable of entering the radar receiver at the time dictated by the position of the radar range tracking gate. If the radar is range tracking the vehicle carrying the ECM system, the ECM signal must be transmitted with a negligible delay after interception of the radar signal at the ECM antenna.

5.2.2. Pulse Range Tracking ECM

Because the radar angle tracking loop operates only on signals that arrive at the radar during the time of the radar's range tracking gate, the objective of an ECM system designed to interfere with radar tracking operation is to move the range tracking gate away from the true target position to a position at which only the jamming signal is present. However, to move the radar range tracking gate, the jamming signal must first enter the radar receiver when the gate is positioned on the true target signal. This implies that the ECM system must initially transmit jamming signals as coincident in time as possible with the signal reflected from the vehicle being protected. As we indicated in the discussion of jamming the range measurement function of search radars, to inject signals prior to the arrival of, or in exact synchronism with, the pulse at the radar, the ECM system is required to rely on second-time-around signals. STAE signals are ineffective when the radar uses carrier frequency or PRF agility.

Furthermore, because of even miniscule delays in the ECM system due to required detection and amplification of the radar intercepted signals, the leading edge of the least delayed signal from the ECM system will arrive at the radar as much as 100 ns after that of the reflected signal. Much effort is being expended in the design of ECM systems to minimize the delay in the TOA at the radar of the ECM signal relative to the echo signal. This effort has been intensified because radar designers, in their antijamming designs, are exploiting this vulnerability in ECM systems. Leading-edge tracker antijamming, discussed in the next section, is one of the radar techniques that exploits this vulnerability in ECM systems.

As we discussed in Section 2.12.1 on frequency modulated pulse compression radars, the processed short-pulse output of that radar is not produced until the last of the long pulse has passed through the compression circuits. Therefore, an interfering pulse that is transmitted, even with the delay experienced in ECM systems, can exit the pulse compression circuit well in advance of the compressed pulse. An ECM signal that is retransmitted as a replica of the radar's extended pulse can therefore be modulated to produce a compressed pulse, which appears at a shorter range than the true reflected signal. This identifies an exploitable vulnerability of frequency modulation pulse compression radars in that the delay in pulse compression allows interfering signals to enter the range and angle tracking loops before arrival of the true echo, which is a very desirable characteristic for ECM systems.

5.2.2.1 Noise

As shown in Figure 5.2, a noise signal can be transmitted by the ECM system in a manner such that it masks the true target signal and the regions on either side (in range) of the true target position. In this case, if the noise signal is sufficiently greater (> 3 dB) than the true target signal, the tendency for the radar range tracking loop is to wander in range and will very likely reach an equilibrium position at one or the other of the edges of the noise burst. Assuming that this phenomenon occurs, the range tracking gate is removed from the true target position to one at which only the ECM signal is present in the angle tracking loop. Appropriate angle jamming modulation of the noise signal will therefore affect the operation of that loop in the radar.

As shown in the figure, the noise signal is based on the premise that neither carrier frequency agility nor PRF agility is used by the victim radar. Otherwise, it would be difficult, if not impossible, to generate a noise signal whose leading edge is at a nearer range to the radar than is the true target. In that case, it is likely that the radar range tracking gate would reach equilibrium at the leading edge of the noise signal at the true target position.

5.2.2.2 Range Gate Pull Off

Assuming that the ECM system can transmit a signal which is almost coincident

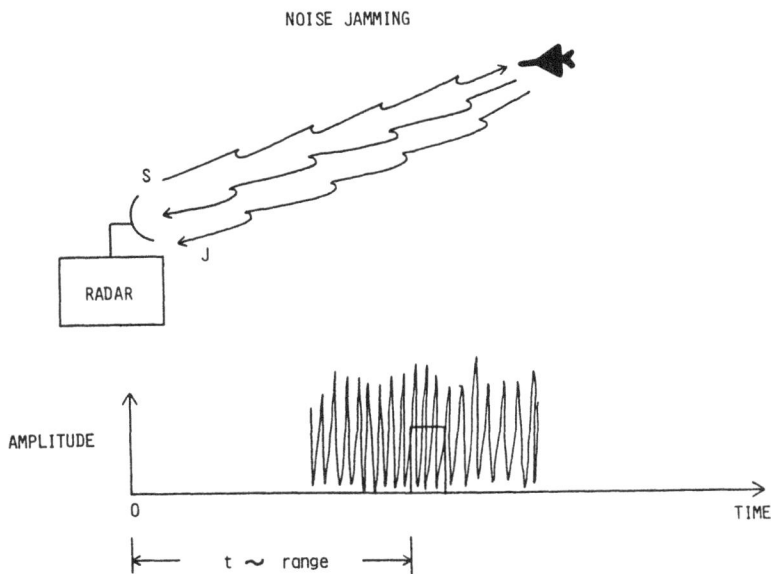

Figure 5.2 Noise range jamming.

with the echo signal, as shown by the solid line pulse in Figure 5.3, the range tracking and angle tracking loops are thus both subjected to the modulations on the ECM signal. Simultaneous arrival of the ECM signal with the echo signal implies no time delay on the ECM signal, resulting in little, if any, disturbance of the target range measurement or tracking: However, as indicated in later chapters, other modulations can be imposed on the signal, which can disturb the angle measurement function of the radar. As stated previously, the angle tracking loop is forced to process all signals that are gated through to it by the radar tracking gate.

However, because (in this case), the true echo signal is also included in the tracking gate and it contains modulations which identify the true angular position of the target, the ECM signal can influence the angle measurement only if it is of sufficient intensity to override the modulation contained on the true target signal. For some radars, when the ECM signal is coincident with the true reflected signal, the J/S requirement for effective angle jamming can be as much as 40 dB.

If the tracking gate can be moved away from the true target position and is forced to operate only on the ECM signal without the true target data, an infinite J/S is achieved because the S in the ratio is equal to zero. This is the technique used by ECM systems with what is known as the *range gate pull-off* (RGPO) technique.

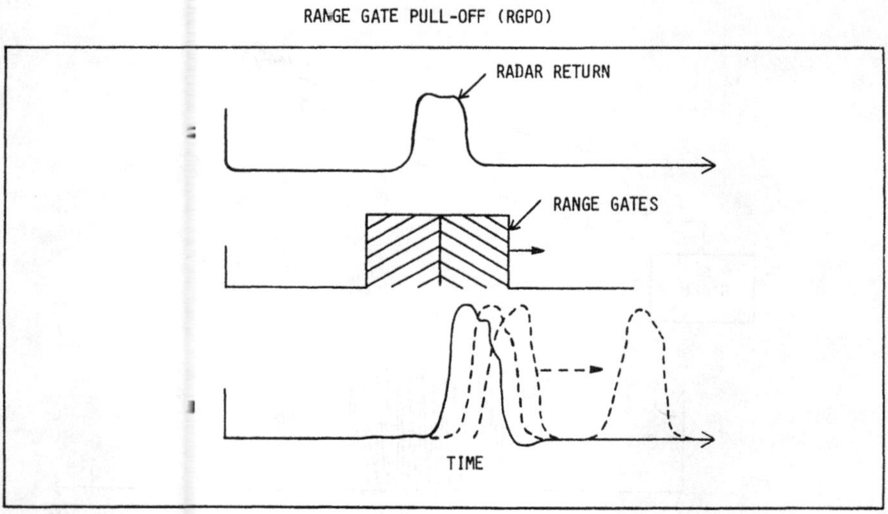

RANGE GATE PULL-OFF (RGPO)

Figure 5.3 Range Gate Pull-Off (RGPO).

As shown in Figure 5.3, the pulses indicated by the dotted lines show subsequent positions of the solid line pulse (the ECM pulse) moving away from the true target position in extremely small steps (on the order of nanoseconds) until its position is at least two or three pulsewidths away from the true target position. If the energy level of the ECM pulse is sufficiently greater than that of the target echo (>3 dB), the range tracking gate is influenced by the ECM signal, and is forced to follow it away from the true target.

Of great importance is that the rate of pull-off be within the design capabilities of the radar tracking loop because ECM pulse movement represents the equivalent of a target moving relative to the true target. If the rate of pull-off is too great, the radar tracking loop may not be able to follow the pull-off target. In that case, the tracking gate will release the false target and fall back onto the true target because of the memory capabilities designed into tracking loops. Typical pull-off rates are on the order of three times the acceleration due to gravity (3 g), which is within the capability of most radar tracking loops designed to operate against military targets.

Figure 5.4 shows a typical RGPO program. The y-coordinate represents the delay (range difference) imposed on the pull-off false target relative to the true target position. The x-coordinate is real time; the time for one pull-off cycle is typically as much as six seconds. As shown at the beginning of any one pull-off cycle, the delay is at a minimum, so that the pull-off false target is almost coincident with the true target pulse position, which is also the position of the radar range

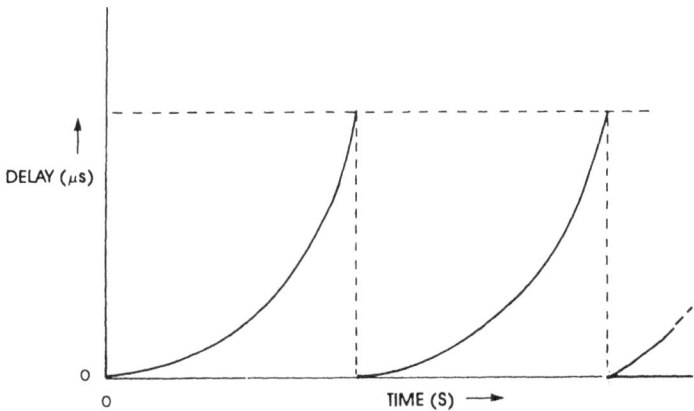

Figure 5.4 Delay program for RGPO.

tracking gate. The degree of coincidence is a function of the minimum delay that can be built into the ECM system; current systems have been designed with as little as 50 ns of delay.

As shown in the figure, the pull-off program is generally parabolic in form, with minimum change in delay at the beginning of the program and maximum toward the end of it. The form of the curve is based on a desired pull-off acceleration level relative to the true target motion. Typically, the pull-off rate is on the order of 3 g. Using the formula $r = 1/2\, a \cdot t^2$, the total delay after 6 seconds is about 2 μs, which is about 4 pulsewidths distant from a pulse at the true target position (assuming a radar pulsewidth of 0.5 μs).

Because the pull-off target is competing with the true target at its minimum position, the radar range tracking loop is influenced by the relative velocity and acceleration characteristics of the true target. Any sudden change in these characteristics in the pull-off program may be ignored by the range tracking loop. Therefore, at the minimum delay position, the rate of change of delay is minimal, with an increasing rate of change as the false target and, perhaps, the radar tracking gate are pulled away from that position. If the pull-off rate is within the design capabilities of the range tracking loop, the tracking gate will be forced to follow the false target, which is the objective of the ECM system.

The rate of change of delay (the slope of the curve shown) represents a change in velocity relative to that which already exists on the true target with respect to the radar position. The rate of change of velocity (the *derivative of velocity*) is equal to the false target's acceleration relative to the true target. We must emphasize that radar range tracking loops are designed with limits in their acceleration capability. Therefore, if the radar is already tracking a target with an acceleration equal to that limit, any further acceleration imposed by a pull-off program will not be accepted by the tracking loop. In this case, the radar tracking loop will tend to remain on the true target. However, in the general case, we can expect that the radar range tracking loop will operate with acceleration levels of somewhat less than the design maximum.

Although as much as 40 dB of J/S is required for an ECM signal to disturb radar angle tracking when it competes with the true reflected signal, J/S levels as low as 3 dB are adequate to produce an RGPO in most radars. Therefore, RGPO is most often used as a "means to an end"; that is, with a 3 dB J/S, the radar range tracking gate may be moved into a position where the angle track loop jamming can be used without competition from the true target, thus effectively being an infinite J/S. The angle jamming modulation is applied to the ECM signal during the period that the radar tracking gate is away from the true target position, after the range gate has been pulled away from the true target position by the RGPO technique.

The pull-off program shown in Figure 5.5 is toward the right of the figure. This is in the same direction as a target moving outward in range from the radar. A program outward in range can easily be generated because these signals are transmitted immediately after the arrival of the pulse at the ECM system. Although

RANGE GATE PULL-OFF (RGPO) PLUS HOOK

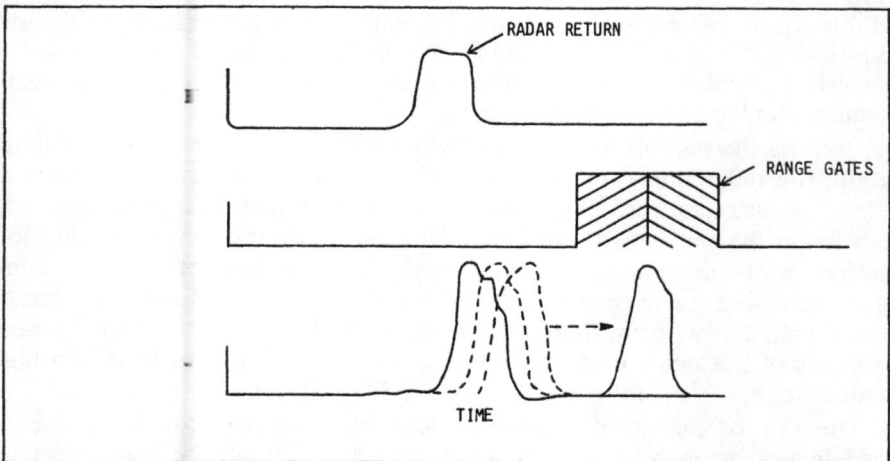

Figure 5.5 RGPO with hold-out target.

the figure may imply that the pull-off targets are generated from a single pulse, in operation, only one delayed pulse is transmitted for each pulse intercepted. At each successive transmission, the ECM pulse is transmitted slightly later in time (on the order of nanoseconds). Typically, the pull-off to about four pulsewidths is completed within six seconds. For example, this would be after 6000 pulses against a radar with a PRF of 1000 pulses/s. Throughout the period when the radar tracking gate is away from the true pulse position, the angle tracking loop is influenced by the modulations on the false target and not the true target. During the period of RGPO, the angle jamming modulations are imposed on the ECM signal.

If the RGPO program were unable to capture the radar range tracking gate for any reason, subsequent ECM transmissions and modulations while the ECM signal was outside the range tracking gate would be fruitless. Because there is no means, at the ECM system, of determining that the RGPO program was successful in capturing the radar range tracking gate, the RGPO program must be recycled from the true target position to make another attempt at gate capture. However, it is entirely possible that the RGPO program was effective, and the ECM modulations were on the verge of jamming the angle tracking when recycling was initiated.

To solve the recycling problem, a modification to the RGPO is used by advanced ECM systems, as shown in Figure 5.5. As described above, a pull-off set of targets are indicated by the dotted-line pulses. Another false target is also shown, positioned at the extreme end (typically four pulsewidths) of the pull-off range. This is mechanized via the ECM system by transmitting two pulses for each pulse intercepted from the victim radar; one of these pulses is programmed to move away from the true target position, as before; the other is continuously positioned at the outer range indicated.

The objective of the two-target program is to pull the radar tracking gate away from the true target position and place it on the outside false target. The outside false target is generally referred to as a *hook*, or *hold-out* target, as its role is to hold the tracking gate away from the true target position, even though the pull-off ECM signal is programmed to recycle to the minimum delay position.

Without the hook, recycling the pull-off may otherwise be performed at an inopportune time during the angle jamming program, which is the primary objective of the ECM system. The angle jamming modulation is continuously applied to the hold-out target so that, when the radar range tracking gate hooks onto it, the radar angle tracking loop is continuously exposed to the erroneous angle modulations. The angle modulations required are described in the next chapter.

5.2.3 Leading-Edge Tracking

Radar designers, cognizant of the inherent delay in ECM systems, have developed a radar range tracking technique which exploits this vulnerability. The technique used is referred to as *leading-edge tracking*. The argument for this technique is that if the radar gates through only the leading edge of the reflected pulse, none of the inherently delayed ECM signal will be gated into the processing circuits of the radar. In effect, the ECM signal is denied the capability of influencing either the AGC, range, or angle tracking circuits.

As shown in Figure 5.6, the incoming signal is differentiated and, as indicated by (b), a signal is present due only to the leading edge of the pulse. In this case, the split gates are just large enough to encompass the leading-edge pulse. The range tracking loop, and thus the angle tracking loop, operate on only that part of the received pulse. The radar tracking gate is terminated before any part of the interfering signal can enter the receiver because of the inherent delay in ECM systems.

The use of the leading-edge tracking gate ECCM technique, however, compromises the performance of the radar because the processing circuits are forced to operate on a much smaller energy level. Furthermore, because of the variable position of the effective center of the radar cross section of the target vehicle, the true target pulse and thus its leading edge can jitter in position by more than the inherent delay in ECM systems. This results in the leading edge of the reflected signal jumping in and out of the radar tracking gates. Therefore, the leading-edge tracking antijamming technique is accomplished only with an attendant degradation in radar tracking performance.

As indicated above, against pulse compression radars, generating an ECM signal that exits the radar processing circuit at a time (range) less than the compressed true reflected signal is possible. In this case, the leading-edge tracking technique is not applicable.

5.2.4 Pulsed Doppler Range Tracking

Although the primary sorting parameter in doppler radars is with the doppler frequency of the return signals, some fire control radars employ range tracking to provide a more accurate measurement of range to the target than is possible with the multitude of fixed contiguous range gates usually associated with this type of radar.

However, because of the narrow bandwidth of doppler filters, tracking the individual pulse returns as is done in conventional pulse radars is not possible. To preserve the characteristics of a pulse for such measurement, a bandwidth of several megacycles per second is required, whereas the doppler detection bandwidth is generally on the order of a few hundred hertz.

Figure 5.7 illustrates the technique used by doppler radars to site the centroid of the energy within a return pulse, and yet be able to pass the data via the narrow-bandwidth doppler filter. As shown at (1) in the figure, a gate is generated which varies its time position about the expected TOA of the pulse; the gate variation

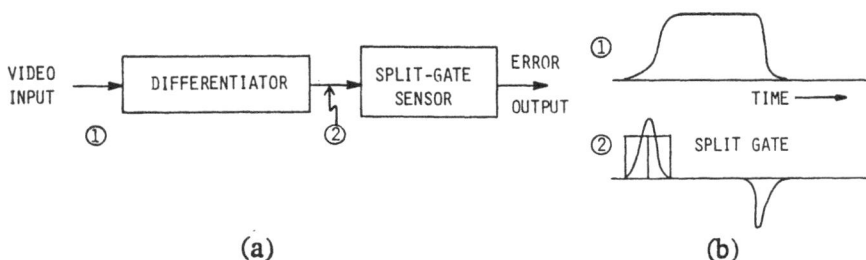

Figure 5.6 Leading-edge tracker.

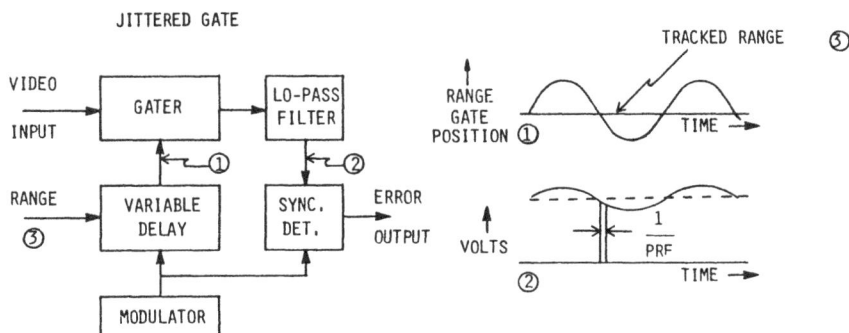

Figure 5.7 Pulsed doppler range tracker.

is performed at a frequency which is less than the doppler filter bandwidth (<1 kHz). The variation in the gate position is as shown at (2). In effect, the variable gate position modulates the incoming pulse so that, if it is not at the correct position, a modulation is produced as an error signal. This error signal is used to adjust the center of the gate variation so that no modulation or error signal results. Because the gate variation is performed at a frequency that is less than the doppler bandwidth, the pulse modulation, and not the pulse, is passed through the narrow-band doppler filter and to the range tracking loop.

5.2.4.1 Pulsed Doppler Range Tracking ECM

The RGPO techniques described previously for the conventional pulse radar are equally applicable to the pulsed doppler type of radar. However, the microwave storage medium required for the time delay modulation of the ECM transmitted signal must be extremely accurate in its frequency reproduction against pulsed doppler radars. This is because the ECM transmitted signal must be within the narrow detection bandwidth of the doppler filter (<1 kHz), which is tracking the true target signal, if interference with that tracking is intended.

Much effort is being expended to provide the type of accurate microwave storage medium required for this application and for other coherent ECM techniques to be discussed later. The ECM transmitted signal is required to be, in effect, an exact replica of the radar signal, except for any intentional modulation placed on the signal for jamming of the radar, be it in time, amplitude, or frequency.

Assuming that a coherent microwave storage system is available, the same type of programmed delay described for conventional pulse RGPO is equally applicable against a coherent radar that employs a range tracking gate to sort signals for the angle tracking loop. However, because the doppler frequency of the target signal is the primary sorting parameter for doppler radars, the value of ECM techniques designed to interfere with the radar's range tracking is questionable.

5.2.5   Radar Doppler Tracking

The primary selection and sorting parameter in a doppler radar, either CW or pulsed doppler, is the doppler frequency of the detected signals. As described in Section 3.2.2, the doppler frequency is directly proportional to the range rate of the target relative to the radar. Doppler filtering of the incoming signals eliminates other signals that may be at the same angle or even within the same range interval. Doppler filtering also minimizes the noise present in the detection process, thereby restricting any interference, intentional or unintentional, to that which exists within the same doppler bandwidth as the true signal. Doppler filtering provides a parameter for identification, selection, or rejection of targets within the range of interest to the radar.

Figure 5.8 shows a typical mechanism for doppler frequency tracking. This mechanism is analogous to the range tracking scheme described earlier in this chapter. Whereas the operation of the range tracker is based on the TOA of the reflected signal, the doppler tracker operates on the basis of its doppler frequency. The two narrow-band filters, (1) and (2), are analogous to the early and late gates of the range tracker; in this case, filter (1) is the "early" filter (as shown in Figure 5.8), and filter (2) is the "late" filter.

Assuming, initially, that the *voltage-controlled oscillator* (VCO) in the figure is properly set in frequency, the target signal from the radar receiver is heterodyned down to the doppler frequency which is exactly in the center of the overlapped filter bandwidths. In this case, the energy level in one filter is exactly equal to that in the other; the error signal at (3) then is at zero, and correction is neither made nor needed.

Hence, if the doppler frequency changes due to relative motion of the tracked

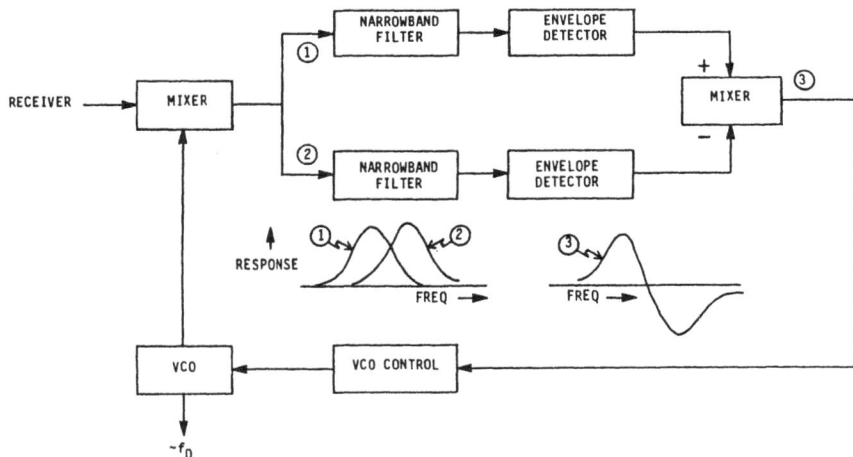

Figure 5.8 Doppler tracking.

target, the energy level in one of the filters (depending on the direction of the change) will be greater than that in the other filter. This difference will produce an error voltage at (3), which adjusts the VCO to minimize the error. This closed loop then continuously corrects the VCO, which helps maintain the error signal at zero. As shown for the range tracking loop, the doppler (velocity) tracking loop is also a closed loop system. In this loop, tracking of the target doppler is performed with a predetermined error, based on the target's velocity and acceleration relative to the radar. As for the range tracking loop, the design of the velocity tracking loop is based on the dynamics of the expected target engagement.

5.2.6 Doppler Tracking ECM

5.2.6.1 Noise

As described in Section 3.3.1.1, there are difficulties in generating noise on both sides of the true target pulse for range track jamming of a conventional pulse radar, primarily because of the carrier frequency and PRF agility of those radars. This is not necessarily the case for doppler radars. They must operate coherently in both parameters (carrier frequency and PRF) to be able to extract the doppler frequency, which is generally on the order of kilohertz. The required coherent time intervals in these radars is on the order of milliseconds, which typically include many pulse interval times of the radar transmission. This requirement forces the radar to be receptive to the same carrier frequency over many pulse intervals, making it vulnerable to second- or multiple-time-around signals, be they intentionally or unintentionally generated.

Furthermore, applying the analogy of the early and late gates of range tracking to the early and late filters of doppler tracking, an ECM system is able to generate signals at doppler frequencies on either side of the true doppler filter to provide interference on both sides of the doppler tracker. As described in the section on search radars, generating doppler frequencies on either side of the true doppler is relatively easy. These doppler-generating techniques include *serrodyning* (transit-time modulation of a TWT) and the technique of phase modulation of the true signal passing through the ECM system amplifiers.

5.2.6.2 Velocity Gate Pull-Off

The diagrams shown in Figure 5.9 for pull-off of the velocity tracking filter are much like those shown in Figure 5.3 for pull-off of the range tracking gate. The only difference is that the x-coordinate in this figure is doppler frequency, whereas it is time (range) in the range tracking gate. The objective of the *velocity gate pull-off* (VGPO) technique is to capture the doppler tracking gate with a sufficiently intense false doppler signal and to move the gate away from the true target position to a position several filter bandwidths away from the true target doppler. Tests have shown that J/S levels of approximately 3 dB are most often adequate to capture the velocity gate away from the true target doppler. As with the RGPO program, the objective of the VGPO is to capture the radar's primary tracking gate (in this case, the doppler frequency), and to move it to a position away from the true target so that the angle modulation imposed on the ECM signal need not compete with the modulation on the true target signal.

As with the range tracking gate, the pull-off rate of the false target must be within the design capabilities of the doppler tracking loop because the pull-off program represents a target that is accelerating relative to the true target. As shown in the figure, because doppler tracking is based on frequency rather than time, the pull-off can be programmed on either side of the true doppler. Also shown is a hold-out, or hook, target to hold the doppler tracking gate at the false position, even though the pull-off program is recycled to initiate a new pull-off. An important advantage with the VGPO technique is that the pull-off and hold-out targets can be placed on either side of the true doppler; the hook on the left side (lower frequency) represents a target with a lower relative velocity, and the hook on the right side (higher frequency) represents a target with a higher relative velocity.

Figure 5.10 shows a typical pull-off program for the VGPO technique. As indicated, the format is similar to the RGPO pull-off program, except that the y-coordinate is doppler frequency (velocity) rather than time delay (range). To

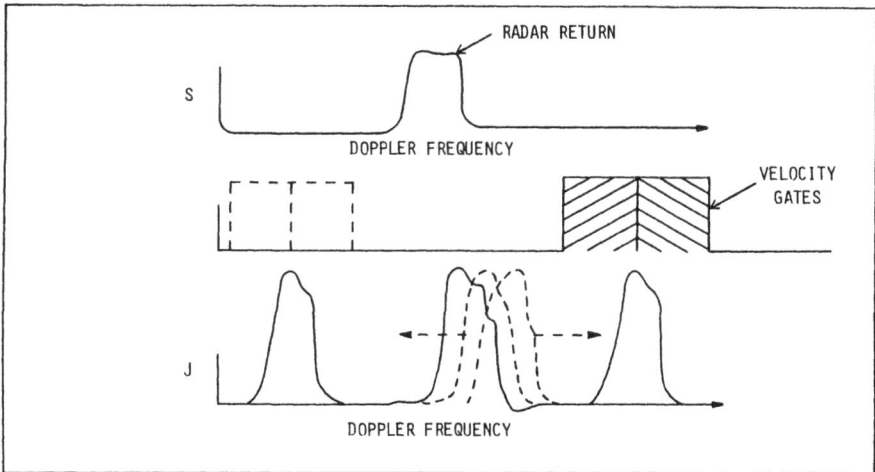

Figure 5.9 VGPO with hold-out target.

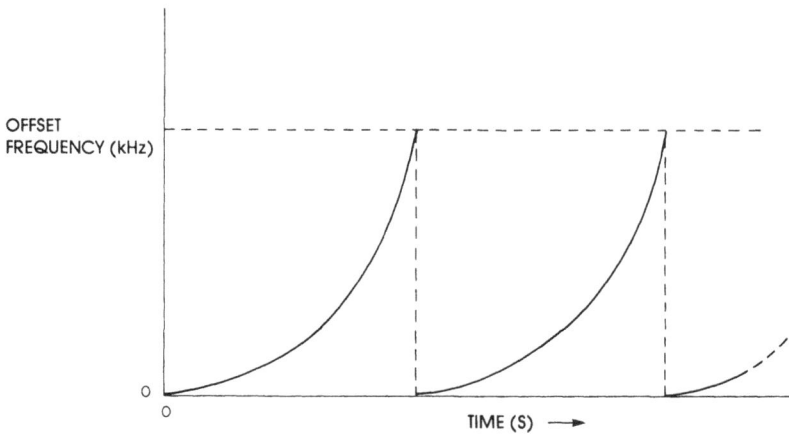

Figure 5.10 Frequency offset program for VGPO.

present a minimum disturbance to the radar velocity tracking loop at the beginning of the pull-off program, a minimum change in doppler frequency is imposed, thereby increasing in rate of change as the false doppler moves away from the true target doppler frequency. Because the rate of change of doppler frequency represents an acceleration, care must be exercised so that the slope of the program does not exceed 3 g, which is typically the design limit of the velocity tracking loops.

In this case, the formula is $v = at$, where the acceleration, a, is now a function of time, t. Using $a = .75\,at$, at $t = 4$ seconds the acceleration is equal to 3 g, which is typically the design maximum of radar velocity tracking loops. The formula for the velocity program then is $v = .375\,gt^2$, which is a parabolic equation providing minimum disturbance at $t = 0$, as the doppler frequency is directly proportional to relative velocity. Therefore, at the end of the pull-off program, the relative velocity of the pull-off false target is 59 m/s (212 km/hr). Translating this to a doppler frequency at a radar carrier frequency of 10 GHz in accordance with Equation (3.2):

$$f\,(\text{Hz}) = 18 \times V\,(\text{km/hr})$$

$$f\,(\text{Hz}) = 3816\,(\text{Hz})$$

Assuming a radar doppler tracking filter width of one kilohertz, this represents a pull-off of almost four filter widths, which should be adequate to remove any influence from the true target.

We must emphasize that the doppler modulation inserted on the ECM signal is an offset to the doppler frequency that already exists on the intercepted signal due to motion of the vehicle relative to the radar. If an acceleration of 3 g already exists on the signal being tracked by the velocity loop, any further acceleration placed on the ECM signal may result in the ECM signal exceeding the dynamic characteristics of the radar velocity tracking loop.

As with the RGPO technique, the main objective of the VGPO technique is to provide a higher J/S for the angle deception techniques. The angle jamming techniques require J/S values as high as 40 dB when competing with the true radar signal, whereas the RGPO and the VGPO techniques only need about 3 dB to be effective. Having captured the tracking gates with a low J/S requirement, the radar tracking gate is moved to a position that includes the false target signal, but not the true target signal, which is effectively an infinite J/S ratio. Because the radar angle tracking loop is influenced only by the signals passing through the tracking gates, it is subjected to the false angle modulations on the ECM signal after VGPO has been accomplished.

Although 3 dB J/S was suggested as necessary to capture the range and velocity gates, we should emphasize that this may not be adequate when an operator is able to observe the movement of the false target relative to the true target on his or her display. Because these gates control signals into the AGC circuits of the radar, a higher J/S (10 dB) may be required to force the AGC to reduce the true target signal level at the display.

Because of microwave signal storage limitations and ECCM considerations, the RGPO technique is limited in its maximum range of pull-off to 2 to 4 μs (only a fraction of a mile in range). Therefore, no significant effect on radar performance can be expected if range-only jamming results. However, because doppler shifts of extensive magnitude (several times the radar PRF, if desired) can be achieved with current ECM systems, a significant effect on radar performance can be achieved with doppler jamming only. This is especially true in the case of interceptors that use doppler frequency measurement in their terminal guidance equations.

5.3 ELECTRONIC COUNTER-COUNTERMEASURES

In the previous discussions which described ECM techniques used to interfere with the radar's range measurement and its range and velocity tracking mechanisms, several radar techniques were suggested to counter those ECM techniques. In this section, we will review ECCM techniques developed and used by radar designers to remove or reduce the effectiveness of the ECM techniques used to attack the range and velocity tracking functions of the radar.

5.3.1 Range Gate Pull-Off ECCM

5.3.1.1 Carrier Frequency and PRF Agility

Several ECCM devices already described in this book as designed to operate against other functions of the radar also have an effect on the effectiveness of the ECM techniques described in this section. For example, the radar carrier frequency and PRF agility capabilities designed into search radars would have the same effect on tracking radars, in effect, the inability to inject signals reliably into the radar receiver to arrive immediately before the leading edge of the true target signal. The net result is that the RGPO technique is generally designed to inject signals into the radar receiver that arrive immediately during or after the arrival of the true target signal, depending on the state of the pull-off program.

5.3.1.2 Leading-Edge Tracking

One of the ECCM techniques developed exclusively to counter the RGPO technique is the *leading-edge tracker*. This technique exploits an inherent vulnerability in ECM systems—the delay incurred in detecting the radar signal and in subsequent transmission of an appropriate response. Minimum delay is achieved when the detected signal itself is used as the carrier for subsequent transmission, as in an ECM repeater. However, delays through the amplifying system and transmission lines can result in as much as 50 to 100 ns of delay between the signal reflected off the vehicle and the least-delayed ECM signal. If the ECM response requires the measurement of the characteristics of the detected radar signal and the subsequent generation of a replica of the signal (as in a transponder), much larger delays (on the order of a 100 ns) must be tolerated.

5.3.1.3 Intrapulse modulation

Until coherent microwave signal storage devices are available, ECM systems must rely on noncoherent storage and reproduction of the jamming signal. Current ECM systems that provide the RGPO capability use noncoherent storage devices for several reasons. The most important reason is that they are fully developed and available at reasonable cost. The second reason is that the technique is primarily used against noncoherent tracking radars, the detection bandwidth of which is on the order of a megahertz. Currently available microwave signal storage devices can reproduce the signal with a power spectrum such that at least 50% of the transmitted energy is within the detection bandwidth of the victim radar. Because of the advent of coherent radars into the hostile environment, much effort is being expended to develop coherent microwave storage systems. One such development is the *digital radio frequency memory* (DRFM) described in Section 9.1.2.

The current generation of technology in noncoherent microwave signal storage devices samples the leading edge of the detected radar signal on the assumption that it is representative of the remainder of the intercepted signal. In many cases, this is a valid (or at least tolerable) assumption. However, this assumption can result in faulty reproduction of the intercepted signal if the latter part of the radar signal is significantly different from that of the leading edge. This is true for radars which provide intrapulse phase or frequency coding of the radar signal. Although it is primarily designed to provide other advantages to the radar, such as *low probability of intercept* (LPI) or high target resolution, intrapulse coding will have a serious effect on proper ECM signal reproduction of the radar signal because of the requirement for coherent microwave signal storage.

A more direct ECCM device used in some radars to counter the RGPO technique generates one or more signals which immediately precede the radar signal used for detection and location of the true target. Only one of the radar signals possesses the true carrier frequency; the others are at frequencies to which the radar receiver is not receptive. This technique is based on the fact that current microwave storage systems operate only on the leading edge of a selected radar-intercepted signal and are blind to any other until as much as 4 µs after detection of the selected signal.

5.3.1.4 Manual Override

Because the RGPO signal is delayed from the true target signal, an experienced and alert operator may be able to differentiate the targets on his or her display. This is one of the reasons that RGPO programs are designed to generate false targets occurring not more than three or four pulsewidths from the true target position. If the display provides a resolution of targets better than this displacement, the operator may still be able to detect the presence of the RGPO false targets. A higher J/S in the ECM signal can capture the radar AGC and suppress the true signal, thus denying operator intervention.

5.3.1.5 Range Guard Gate

A technique suggested for use against the RGPO program is shown in Figure 5.11. A second gate is positioned by the radar immediately adjacent to the radar range tracking gate. The detection of signals in both gates alerts the radar range tracking circuits that a false target program is being attempted against the tracking loop. Appropriate processing is then initiated to maintain track on the signal in the earlier of the two gates.

5.3.2 Velocity Gate Pull-Off ECCM

The requirement for doppler radars to operate coherently over an extended period of time makes them extremely vulnerable to countermeasures to which noncoherent radars may be immune. For example, doppler radars cannot operate with pulse-to-pulse frequency agility nor PRF agility because of the requirement for coherent operation over at least a specified interval of time (on the order of milliseconds). Furthermore, because doppler radars are required to operate against closing (positive doppler) and opening (negative doppler) targets, they cannot eliminate targets, true or false, based on the "polarity" of the doppler frequency. To eliminate or minimize these vulnerabilities, much effort has been expended on developing ECCM techniques for doppler radars.

5.3.2.1 Coherency Check

Current devices used to modulate the ECM signal for the purpose of capturing the radar velocity gate inherently and unintentionally generate spurious signals in addition to the desired signal. This includes phase modulation of the coherent signal passing through the ECM system amplifiers as well as transit-time modulation (serrodyning) of TWTs. These spurious signals are of a high enough energy level to be recognized by detectors positioned at doppler frequency filters in the neighborhood of the filter containing the true target doppler frequency. Much effort is being expended by ECM designers to reduce the spurious signals so as to remove this vulnerability.

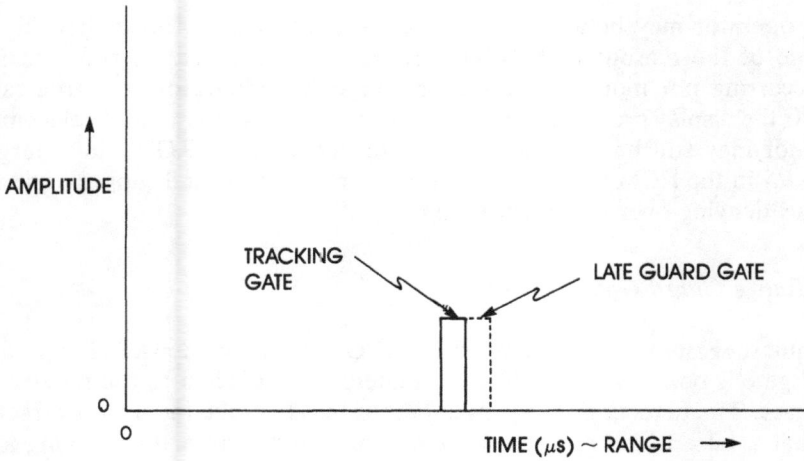

Figure 5.11 Radar guard gates *versus* RGPO.

5.3.2.2   Velocity Guard Gates

As we suggested for the RGPO program, radar designers are using doppler guard filters on either side of the true target doppler filter. As with the range gates, these filters are designed to monitor activity on either side of the true doppler to alert the tracking circuits when VGPO is being attempted.

5.4   MULTIPLE RADAR COMPATIBILITY

Although the description of the ECM techniques in this section was primarily based on a "one-on-one" situation, the designer must consider the ability of the techniques to operate simultaneously against several such radars that may be in the environment and pose a threat to the vehicle being protected.

   Modern ECM systems are being designed with the ability to separate and sort threat signals on the basis of any one of several radar parameters. Whether there is any conflict in operation of ECM techniques when transmitting against more than one radar is important to consider, even when such sorting has been accomplished. Radar parameters primarily used for sorting by ECM systems are TOA, AOA, and carrier frequency of the intercepted signal. Other radar parameters are sometimes used for sorting and separation (ie., pulsewidth, PRF and polarization). These, however, are used primarily for sorting for analysis and identification of signals, and are not necessarily useful in identifying the signals which can be modulated independently in the ECM transmitter. Measurement of PRF, however, is necessary to predict TOA of selected emitters.

5.4.1 Range Gate Pull-Off

As described above for the RGPO program, even when hold-out targets are employed, the ECM system is not occupied with any one radar for more than 4 or 5 μs after interception of any one pulse (i.e, making it available for a new pulse from another radar). Because of the recycling feature of the RGPO program, the program need not begin at minimum delay for each new radar entering the ECM system. The same delay program being used against any one radar can be applied equally against all such tracking radars which are simultaneously in the environment. The microwave storage system used to time modulate the ECM transmitted signal must be capable of storing signals at any of the carrier frequencies of the victim radars in the environment.

If separate and unique programs are required for simultaneous effectiveness against more than one radar, the primary sorting parameters, TOA or carrier frequency, can be used to separate the signals in the ECM system and to apply individual modulations to each. This may be required when a high duty cycle radar, such as a pulsed doppler radar, is also present. However, the high duty cycle of operation of these radars can occupy the ECM system to such a degree that it degrades the system's performance against whatever other coherent or noncoherent tracking radars that may be in the environment.

5.4.2 Velocity Gate Pull-Off

Unlike the RGPO program which occupies the ECM system for a period of time, the VGPO program is an instantaneous phase (frequency) modulation of the signals passing through the amplifiers. Because the frequency modulation imposes a shift in frequency relative to that which is already present on the signal, all signals receive the same doppler offset regardless of the carrier or doppler frequency of the intercepted signal.

As with the RGPO program, the recycling feature of the VGPO program removes the requirement that the stealing program begin at zero doppler frequency offset for each new radar encountered.

5.4.3 Compatibility Between RGPO and VGPO

The RGPO program is a time modulation of the signal passing through the ECM system, whereas the VGPO is a frequency modulation. Both programs may be applied simultaneously, even if coherent doppler radars are not in the environment. The small frequency shifts applied for doppler offsets (on the order of kilohertz), will affect the spectrum only insignificantly because of the wide detection bandwidth of noncoherent radars (on the order of megahertz).

5.5 SUMMARY

Tracking radars are considered to be a direct threat to the survivability of a friendly vehicle intent on penetrating enemy territory. This is because the tracking radar's responsibility is to measure the target parameters with sufficient accuracy so that effective interdiction of the penetrating vehicle can be made by its interceptors, either manned or unmanned. Furthermore, tracking radars are generally used to protect high value targets, so they become a threat at the most crucial point in the mission, the terminal or attack phase.

The tracking radars use either the TOA of the reflected signal or its doppler frequency, in the case of a coherent radar, to separate the targets of interest from other signals which enter the radar antenna. The separation is required to ensure that the radar angle track loop is operating only on the signals of interest.

This chapter described the methods of operation of the range and velocity tracking loops as well as their vulnerability to outside interference. As indicated previously, ECM activity against the range or velocity tracking loops is provided primarily to obtain a J/S advantage for the ECM signal so that effective jamming of the radar angle measurement can be achieved. Whereas as much as 40 dB of J/S is required for effective angle jamming, as little as 3 dB of J/S is adequate to capture either the range or velocity tracking gate. Having achieved gate capture, the ECM signal enjoys an infinite J/S, thereby ensuring effective angle jamming of the radar.

Chapter 6
Tracking Radar Angle Measurement

6.1 INTRODUCTION

Of all the target position parameters measured in radar systems, including early warning and acquisition radars, the angle measurement made by tracking radars is the most significant in providing effective interception of intruders into the defended territories. Other radars and sensors are primarily responsible for identifying a small region where potential targets are located to minimize the search problem for the tracking radars. The target location measurements made by the tracking radars must be sufficiently accurate for properly guiding manned or unmanned interceptors to their targets. In many cases, even if the tracking radar angle measurement is the only position parameter available during the penetration profile, effective interception can still be achieved, albeit with somewhat reduced performance. Without some degree of angular measurement of the target position, the probability of effective interception of the target is nearly impossible.

Therefore, because of the importance of the measurement of this parameter by the tracking radar, much effort has been expended by radar designers to protect this measurement capability, especially in the case of extreme interference, intentional or unintentional. For the same reason, much effort has been expended by ECM system designers to deny radars an angle measurement capability by developing systems and techniques to produce effective interference in the radar's angle measurement circuits. In many cases, ECM against other target parameter measurements by the radar is designed to facilitate the target angle jamming modes of ECM systems. The range and velocity gate pull-off programs discussed in the previous chapter are such ECM techniques.

In this chapter, all forms of angle measurement and tracking by tracking radars are discussed; these include radars using sequential lobing and monopulse techniques. The defense system radars which fall into the category of tracking radars are fire control radars and missile guidance radars. Active and semiactive radars used for terminal guidance in manned or unmanned interceptors are discussed in the next chapter.

6.2 SEQUENTIAL LOBING TRACKING RADARS

Sequential lobing radars are those which nutate (rotate) their antenna gain patterns in such a manner that a modulation is imposed on all signals entering the antenna which are reflections from targets illuminated by the radar. The extent of this modulation as a function of displacement from the antenna boresight is predetermined, and is used to define where in the antenna pattern the signal source is located. The nutation in the antenna pattern is achieved by mechanically rotating the radar antenna, or by using a fixed antenna with electrical nutation of the antenna mainbeam pattern. The latter mechanization is being used in more advanced radars because of the need for multiple target angle tracking and because of the availability of high speed electronically steerable antennas. The techniques of radar measurement used (as well as the ECM techniques suggested) are only secondarily affected by the mechanization used.

The simplest mechanization used by sequential lobing radars is shown in Figure 6.1. Because these radars are most often required to measure the angle to the target in two angular dimensions (azimuth and elevation), an antenna pattern which is narrow in both dimensions is used. Antenna patterns which are narrow in both dimensions are referred to as *pencil beams*. The locus of half-power points on the antenna pattern in space is represented as a circle; this is shown as the solid line at the target position. In the figure, this circle appears as an ellipse because of the perspective nature of the figure. The distance (in this case, the diameter of the circle) between half-power points is the measure of the beamwidth of the antenna pattern. The dotted circle, whose center is at the target, is the locus of the peak of the nutated antenna beam. The other two circles illustrate the position of the pencil beam at two different times of the nutation.

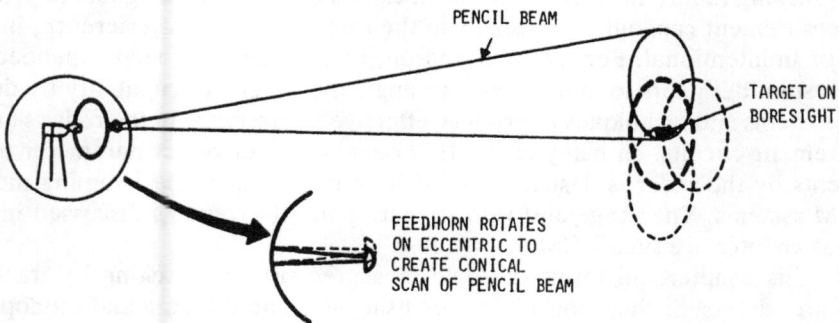

PENCIL BEAM

TARGET ON BORESIGHT

FEEDHORN ROTATES ON ECCENTRIC TO CREATE CONICAL SCAN OF PENCIL BEAM

Figure 6.1 Sequential lobing (conical scan).

In the case shown, the beam is nutated continuously, usually at low frequency (20 to 300 cycles/s), and if the reflecting target is in a position (as shown) on the same point of the pattern no matter where the pencil beam is pointing in its nutation, the antenna is considered to be properly boresighted on the target. If the antenna is pointed off the target, a modulation of the return signal occurs, the modulation indicating the extent and angle of the pointing error. In an angle tracking radar, the amplitude modulation of the reflected signal represents the error signal, and is used to correct the pointing angle of the antenna in order to minimize, if not eliminate, this error.

Figure 6.2 presents a head-on view of the problem, which may be easier to understand. On the left of the figure we can see that the pencil beam is shown as the circle of the half-power points of the antenna pattern, with the dashed circle representing the locus of the peak of the beam as the antenna nutates. We can see in Figure 6.2(a) that if the antenna is indeed boresighted on the target, the amplitude of the return signal is constant. If instead, as shown in Figure 4.2(b), the antenna is pointed off the target, we can see that when the antenna is nutated into position (1), the target is nearer to the peak of the beam than it is when the antenna is nutated to position (2). This results in a modulation of the return signal as shown in Figure 6.2(b); the peak of the wave occurs when the target is nearest the peak of the nutated beam and the minimum occurs when the target is farthest from the peak.

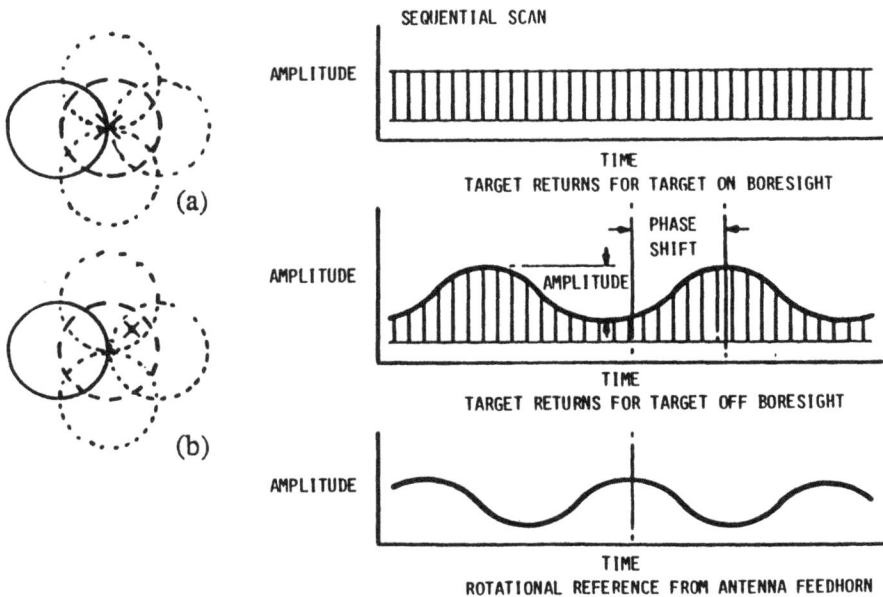

Figure 6.2 Amplitude modulation with conical scan.

As shown, the modulation is at the same frequency as the reference signal generated by the nutating antenna (Figure 6.2(c)), except that the phase of the modulation is dependent on the direction of the target relative to a reference direction predetermined by the radar. Typically, the amplitude of the signal from the reference generator is at positive maximum when the antenna is nutated to the maximum elevation position; the reference signal amplitude is at negative maximum when the antenna is nutated to the minimum elevation position. A target located at the maximum elevation relative to boresight will then produce a modulation the maximum of which will occur at the same time as the maximum on the reference signal. A target at the maximum right azimuth will produce a modulation the maximum of which occurs with a phase shift of 90° relative to the modulation on the reference signal. Taking it one step further, a target at the lowest elevation position of the nutating antenna will produce a modulation the maximum of which occurs 180° removed from the reference signal maximum.

The percentage of modulation of the error signal is dependent on the distance of the target from the boresight position, the maximum occurring when the target is about a half beamwidth from boresight or on the locus of the center of the nutating beam. If the target is on the boresight of the nutating antenna pattern, no modulation of the return signal exists. *Boresight* of the nutating antenna is defined as the center of the circle which represents the locus of the peak of the nutating beam.

The crossover point of a nutating antenna is defined as the point on the antenna gain pattern located at the center of the circle traced out by the locus of the center of the nutating beam. Although the locus of the center of the antenna beam in Figure 6.2(c) is chosen so that the crossover point of the nutation is at the half-power point of the antenna, generally the crossover point is chosen to be nearer to the peak of the antenna, but far enough off the mainbeam so that an adequate modulation of the off-boresight target is present. Furthermore, the illustration assumes that the crossover is at the half-power point, whether the pattern shown is due to a one-way antenna pattern (receiving only) or a two-way antenna pattern (transmitting and receiving). This assumption does not affect the discussions in this chapter.

Figure 6.3 presents a simplified diagram of a typical sequential angle tracking radar. As indicated previously, we assume that the search or acquisition radar or other sensor defines the approximate angular position of the target for initial positioning of the antenna. The radar transmitter (pulse, CW, or pulsed doppler), emits via the antenna, which may be nutating in transmission as well as receiving, depending on the design of the antenna. The nutator generates a reference signal, which is compared in the *error detector* with the modulation on the return signal received via the *transmitting receiving switch* (T/R) and radar receiver. No modulation on the return signal indicates that the antenna is properly boresighted on the target.

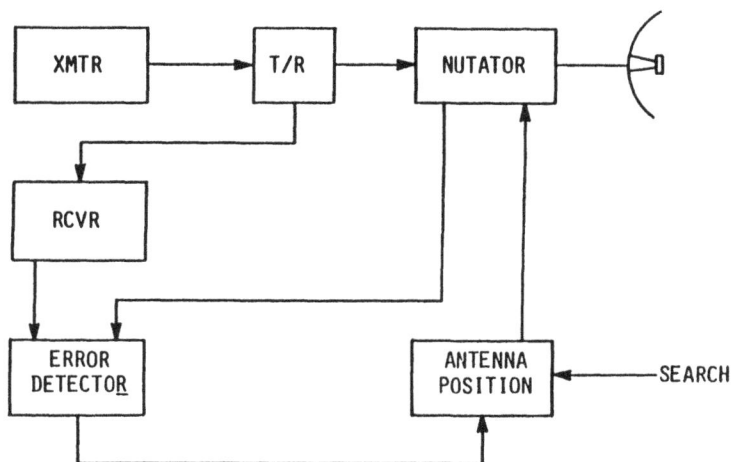

Figure 6.3 Sequential scan angle tracker.

Modulation on the return signal provides the antenna positioning circuit data as to which direction to correct and to what extent; the phase of the modulation indicates the direction of the error, and the amplitude indicates the distance from boresight. This is performed smoothly in a closed-loop circuit until a null is achieved in the modulation, indicating proper boresighting of the nutating antenna. As the target moves in angle, the loop continues to correct, maintaining minimum modulation on the reflected signal. As with the range tracking loop, the angle tracking loop is designed to track only expected realistic angular motion of its targets. The loop is also designed to provide a memory capability, which allows the tracking loop to coast through periods of no data caused by absence of the signal due to fades or other interference.

Although Figure 6.3 is based on an antenna nutating in a continuous fashion about the target position, the nutating program can be of any variety for which a reference signal can be generated. One such program may be an antenna beam that is switched randomly from one position to another about the antenna boresight. As long as this program, though random, can be stored as a reference for the return signal, there is no compromise to the tracking capability of the system. As we will discuss later, such programs may be used by radar designers to counter ECM systems.

The angle tracking loop is generally designed to operate on one return signal at any one time. As a result, the receiver of the radar must be able to dictate which, among all of the return signals present, is the target of interest, even when the antenna is pointed at only one particular angle. Because of the availability of high speed processing circuits and steerable antennas, more advanced radars can maintain angle track on several targets simultaneously. In any case, for proper operation against any of the targets, presorting of the signals is required to separate the modulation data on each signal before insertion into the angle tracking loops.

Even though multiple targets may be contained in the same radar antenna beamwidth, they may be at other ranges or doppler frequencies. Conventional pulse radars sort out these signals in their range tracking circuits before gating them into the angle tracking loop; doppler radars sort out the signals in their velocity (doppler) tracking circuits. (This process has been described in Chapter 5.) In this manner, the angle tracking loop is influenced only by the signal passed through the range or velocity tracking gates, whether it is the true signal or a false signal. Angle tracking loops generally do not possess circuits to determine the authenticity of the signals; that responsibility is left to the detection, identification and gating functions of the radar.

6.2.1 ECM *versus* Sequential Lobing Radars

To attack the radar angle tracking loop, we must first ensure that the ECM signal is passed through the radar range or velocity tracking gate; this is accomplished as discussed in Chapter 5. Having accomplished that, we must then place the appropriate modulation on the ECM signal which the angle tracking loop will accept and process. Of immediate importance is that the modulation must be as near as possible to the nutation frequency used by the radar. This requirement is because, typically, the acceptance bandwidth of the radar angle tracking loop is only about 1 to 2 Hz, and is centered at the reference frequency. Amplitude modulation outside this bandwidth will have no degrading effect on the operation of the angle tracking loop.

Even though the true reflected signal may be overwhelmed by the ECM signal, unless the modulation frequency is within the acceptance bandwidth of the loop, the radar angle tracking loop will coast on its prior data or extract the proper modulation that may be contained on the ECM signal. True target position modulation may exist on the ECM signal because the nutation of the radar receiving antenna will modulate all received signals (true or false), even noise interference. Because of this possibility, the false modulation must be placed on the ECM signal with a high percentage of modulation relative to that imposed on the signal by the nutating radar antenna.

With older radars, which nutated the transmitting antenna as well as the receiving antenna, it was often possible to detect and measure the radar antenna nutation frequency at the ECM receiver. Because the radar angle tracking loop most often tracks with a small error (which is inherent in the type of loop used), the antenna is pointed slightly off the target position, even when appropriately boresighted. The result is that the received power level at the ECM system is amplitude modulated at the same frequency as the reflected signal, due to the nutating radar transmitting antenna.

With this detected data on the radar nutation frequency, the ECM system can then amplitude modulate transmitted false targets or noise at that frequency, but with a phase shift relative to that detected at the ECM receiver. Because the frequency of the modulation on the radar transmitted signal (as intercepted at the ECM receiver) is the same as the frequency of the reference signal in the radar angle tracking loop, it is possible to amplitude modulate the ECM transmitted signal with a different phase and increased amplitude to produce an erroneous error signal in the angle tracking loop.

The optimum phase on the amplitude modulation of the ECM signal is one which is shifted in phase 180° relative to the detected phase. Whereas the true error signal is at a phase and amplitude necessary to correct the position of the tracking radar antenna in the direction of the target position, a modulation 180° out of phase will force the antenna away from the target rather than toward it.

An effective way for an ECM system to accomplish the required modulation is to amplify the detected amplitude modulation at the ECM receiver, invert it (180° phase shift), and use it to amplitude modulate the ECM transmitted signal. The modulation generally used is a square wave at the nutation frequency, which is easily produced with an "on-and-off" gate. The use of a square wave for the amplitude modulation of the ECM signal produces a high percentage of modulation, which is required to compete with the inherent correct modulation also contained on the ECM signal. The square wave modulation also produces many harmonics of the radar nutation frequency, which provides the ECM system an advantage to be described later.

The result of this type of technique, referred to as the *inverse gain* technique, is shown in Figure 6.4. Figure 6.4(a) shows the modulation detected at the ECM receiver; Figure 6.4(b) shows the inverse (180° shift) square wave modulation of the transmitted signal. Figure 6.4(c) displays the pulse signals and the resultant modulation on the signals passed through to the angle tracking loop. As we could see in Figure 6.4(c), the true target signal was also present, as it would be if no RGPO or VGPO process were used to remove the competition of the true target. We can see that the true modulation is present on both the false signal and the true signal. If the J/S or the percentage of modulation of the false signal relative to the true signal is not adequate, the angle tracking loop can still properly track the true target angular position.

Figure 6.4(b) shows the result in the radar angle tracking loop if RGPO or VGPO was first successfully applied in the ECM process; in this case, the true target signal is not present. Also, the true modulation is minimal compared to the square wave modulation on the received false target signals. With this condition, the angle tracking loop is forced to move away from the target rather than toward it, which is the objective of the ECM technique.

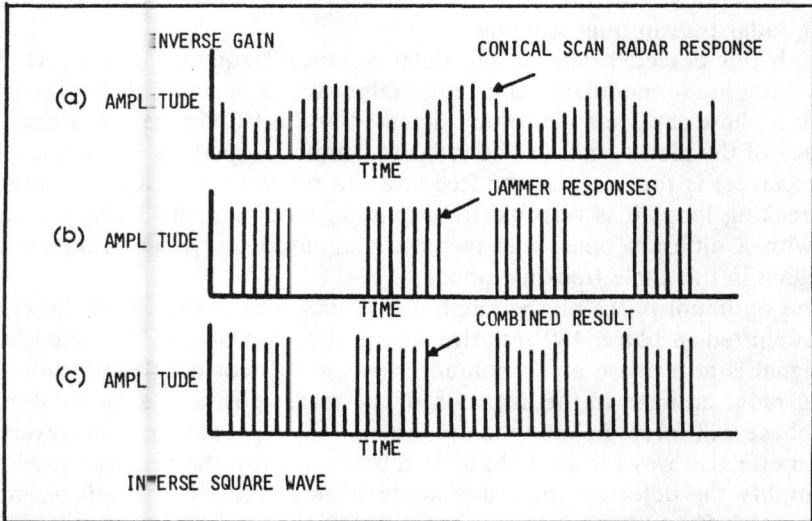

Figure 6.4 Inverse gain on pulses.

6.2.1.1 Use of Auxiliary Antenna

A technique used by active sequential lobing radar systems to counter the effectiveness of the inverse gain technique is shown in Figure 6.5. This mechanization is akin to that described in Chapter 5 for sidelobe blanking techniques (an auxiliary antenna is used to complement the main antenna). In the case of search radar this mechanization was used to identify and cancel signals which were injected into the sidelobes of the radar antenna. In the case of tracking radar, the objective of the two-antenna technique is to remove all external amplitude modulations on the intercepted signal (including any jamming modulations transmitted by ECM systems) from the detection circuits of the radar receiver.

The concept used with this mechanization is that the A antenna provides the radar receiving antenna pattern, which is nutated as required for accurate angle tracking. However, the B antenna is not nutated. As a result, the signals entering the A receiver contain all amplitude modulations (those due to external influences and those imposed by the nutating receiving antenna). The external influences include the modulation due to the nutation of the radar transmitting antenna (if it is actively scanning), and any modulation of externally transmitted sources such as an inverse-gain ECM system.

Conversely, the signals received via the auxiliary antenna contain only external modulations, because this antenna is not nutated. A subtraction process on

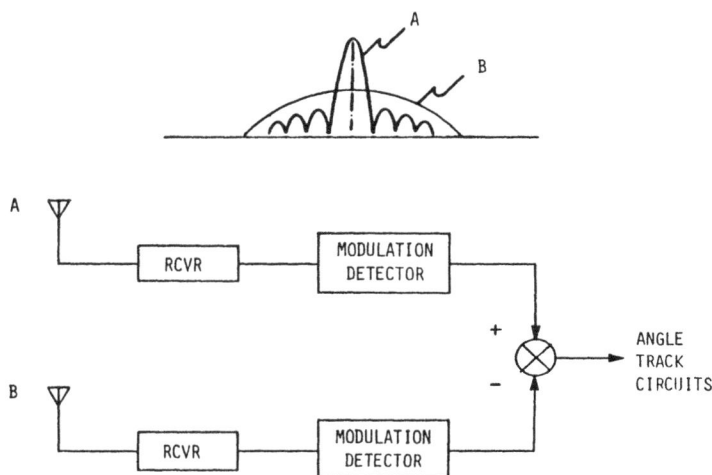

Figure 6.5 Guard antenna mechanization.

the A and B receiver outputs produces a signal output which contains only modulation due to the nutating radar receiving antenna. This process removes all externally generated amplitude modulation on any signals intercepted by these antennas, including the ECM signals. Because the modulation on the received signals due to the nutating receive antenna is contained only in the A receiver, only that modulation appears at the output of the subtracting process.

6.2.1.2 Lobe On Receive Only Radars

To counter inverse gain-type ECM systems, radar designers have resorted to the use of antennas which nutate only during the receiving function of the radar system. Generally, this is accomplished with the use of two antennas: one, the *transmitting antenna,* does not nutate; the other, the *receiving antenna,* does nutate to produce the amplitude modulation required on the reflected signal for effective angle tracking. Radar antennas have also been designed to cause nutation only during the receiving function, even though the same antenna is used for both the transmitting and receiving functions. These radars are known as *lobe on receive only* (LORO) or *conical scan on receive only* (COSRO) radars.

Because the radar transmitting antenna is not nutated in the LORO technique, the ECM system is not able to detect the modulation required for effective inverse gain modulation. It has been determined, however, that, even with the

LORO antenna, a measurable modulation can still be detected because of crosstalk in these antennas between the transmitting and the nutating receiver feeds. If this modulation cannot be detected, alternate techniques of ECM are required.

6.2.2 Swept Square Wave ECM

A technique used by ECM system designers to overcome the inability to detect and measure the radar nutation frequency when engaged with a LORO radar is known as the *swept square wave* (SSW) technique. The concept is shown in Figure 6.6. In this case, the frequency of the amplitude modulation on the ECM signal is varied continuously over the expected range of nutation frequencies which may be used by the radars of interest. This range is usually based on *a priori* data obtained in previous contacts with these radars. The frequency of the amplitude modulation of the ECM transmitted signal is swept from the expected low nutation frequency to the expected high nutation frequency. As indicated by the dotted line which shows the location of a typical nutation frequency, the amplitude modulation on the ECM signal eventually crosses the proper frequency and, if it dwells at that frequency long enough, disturbance of the radar angle tracking loop should result.

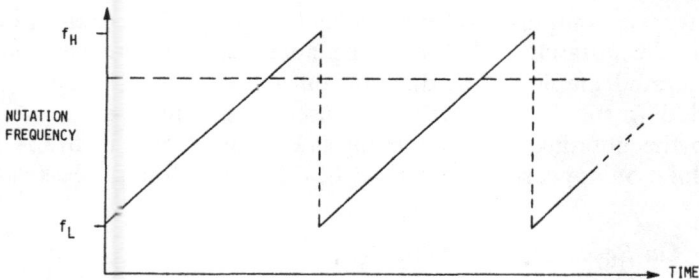

Figure 6.6 Swept square wave.

It is important, however, that the sweep rate be slow enough so that the amplitude modulation exists at the proper frequency for a sufficient length of time so that disruption of the angle tracking loop results. Because the angle tracking loop bandwidth is typically very narrow (on the order of 1 Hz), it is imperative that the frequency of the ECM signal amplitude modulation exist within the loop bandwidth for at least one second. Assuming a frequency sweep range of 100 Hz, it is possible that at least 100 s can elapse before the amplitude modulation reaches the proper frequency. Even with a frequency sweep range of 4 Hz, a four-second block of angle tracking without interference is available to the radar. This may be enough to allow effective interception by the radar defense system, depending on the condition of engagement at the time.

When very large sweep widths are required with the SSW technique, ECM designers have taken advantage of the harmonics generated with square wave amplitude modulation of the ECM signal; a square wave is rich in odd harmonics of the modulation frequency. For example, with a fundamental square wave sweep of 20 to 60 Hz, a third harmonic exists at 60 to 180 Hz, and a fifth harmonic exists at 100 to 300 Hz. Therefore, with a single sweep of a square wave from 20 to 60 Hz, modulations exist simultaneously from 20 to 300 Hz, albeit with reduced amplitude at the harmonic frequencies.

One method of reducing the "free look" time to the radar is to reduce the modulation sweep range to one which more closely approximates the expected region of the radar's nutation frequency. Another method is to increase the frequency sweep rate. However, increasing the sweep rate is accompanied by the probability that the modulation frequency will not be within the tracking loop bandwidth long enough to cause adequate disturbance of the angle tracker.

6.2.2.1   Jog Detection

To overcome the frequency sweep rate limitation, more advanced ECM systems have been designed with a technique known as *jog detection,* which is used with the SSW technique. With this technique, it is possible to increase the sweep speed by as much as a factor of ten over the expected frequency range of unknown nutation frequencies. This technique is based on the fact that, even though the sweep speed is such that the ECM modulation does not dwell long enough to effectively disrupt the radar angle tracking function, it perturbs the tracking loop. Indeed, after the sweeping modulation frequency passes out of the tracking loop bandwidth, the angle tracking loop recovers from the perturbation and resumes proper tracking.

This small movement of the tracking antenna has been shown to cause an amplitude modulation of the intercepted radar signal which can be detected by a properly designed receiver in the ECM system. Because the characteristics of this perturbation are unique and predictable, fairly reliable detection of the time of the disturbance can be accomplished. When the perturbation caused by the sweeping frequency is detected, the ECM signal modulation sweeps over a very small portion of the sweep range at a frequency defined by the time of detection of the perturbation at the ECM system.

The jog detection process is shown in Figure 6.7, where the second sweep caused a perturbation in the ECM receiver detection circuits as it passed through the location of the radar nutation frequency. Continuous modulation of the ECM signal about that detected frequency eventually drives the tracking loop off the target, as shown by the loss of radar signal in the jog detection receiver.

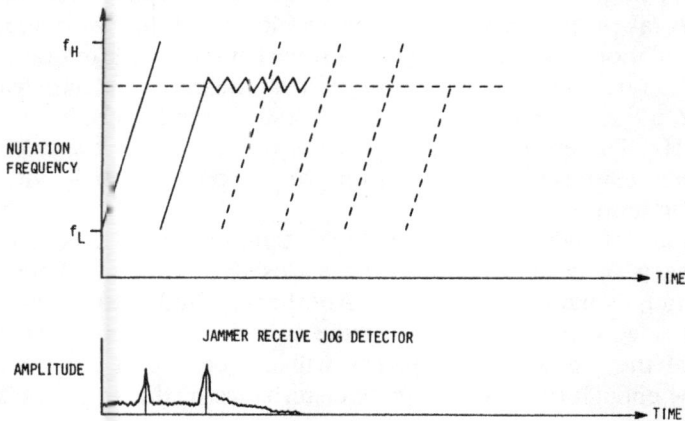

Figure 6.7 Swept square wave with jog detection.

6.2.3 *J/S* Required

In tests, we have seen that the amount of Jamming-to-Signal ratio required to effectively employ the inverse gain or SSW angle jamming technique is 10 to 20 dB when the ECM signal must compete with the true target return (when the ECM signal and the true target are in the same range or velocity tracking gate). However, if the RGPO or VGPO programs are successfully employed to move the range or velocity tracking gate away from the true target position, 3 dB of *J/S* is adequate. Having captured the radar range or velocity tracking gate, the ECM signal enjoys an infinite *J/S* for effective angle jamming.

The techniques discussed in this section are equally applicable against any of the tracking radars using sequential lobing (e.g., pulse, FMCW, or pulsed doppler). The amplitude modulation required for the techniques is imposed on any of the ECM signals, whether coherently repeated or reproduced with microwave storage memories.

6.3 MONOPULSE TRACKING RADARS

6.3.1 Introduction

Whereas a sequential lobing tracking radar must rely on a continuum of return pulses to extract the amplitude modulation which defines its antenna boresight position relative to the target, a monopulse radar theoretically can extract the needed information from a single-pulse return. Although monopulse radars can theoretically measure angle on a single pulse, most are implemented with AGC loops and similar schemes, which can be vulnerable to time-varying jammer techniques. One such jamming technique is discussed later in this chapter.

To provide single-pulse angle measurement capability, monopulse radars can operate on the amplitude difference or the phase difference of the signal as it is received on two or more displaced elements of a radar antenna.

6.3.2 Amplitude Monopulse Tracking Radar

A simplified diagram of a typical amplitude monopulse radar is shown in Figure 6.8. The Δ signal in the figure is the difference in amplitude due to the position of the signal source relative to the antenna boresight in azimuth. The block diagram for only one dimension (azimuth) is shown in the figure; the other dimension is an exact duplicate of this block diagram, except that the Δ signal out of the antenna is due to the target position in elevation relative to antenna boresight. As shown in Figure 6.9, the antenna gain patterns in this one dimension are displaced in azimuth. Because it is the difference signal which identifies the location of the signal source relative to antenna boresight, the antenna difference patterns are shown in Figure 6.9(b). The error voltage associated with these patterns is shown in Figure 6.9(c).

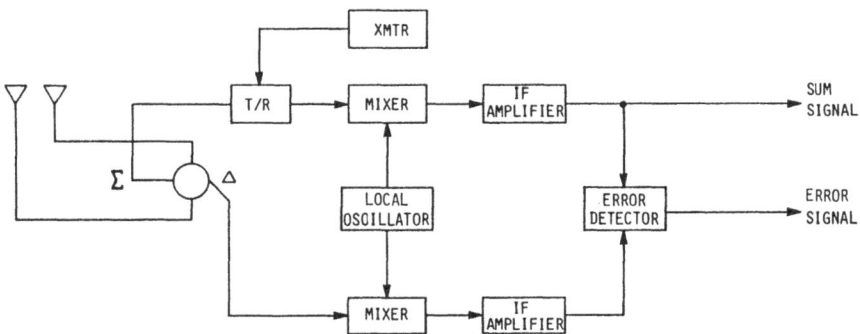

Figure 6.8 Monopulse radar diagram.

Because the error signal is used to correct the boresight position of the radar antenna via the angle tracking loop, unless the error signal is zero, a correction is made to move the antenna beam to produce a zero error. If the error signal is positive, the antenna is moved in one direction toward the proper boresight position; if it is negative, the antenna is moved in the other direction.

GAIN PATTERNS   DIFFERENCE PATTERNS

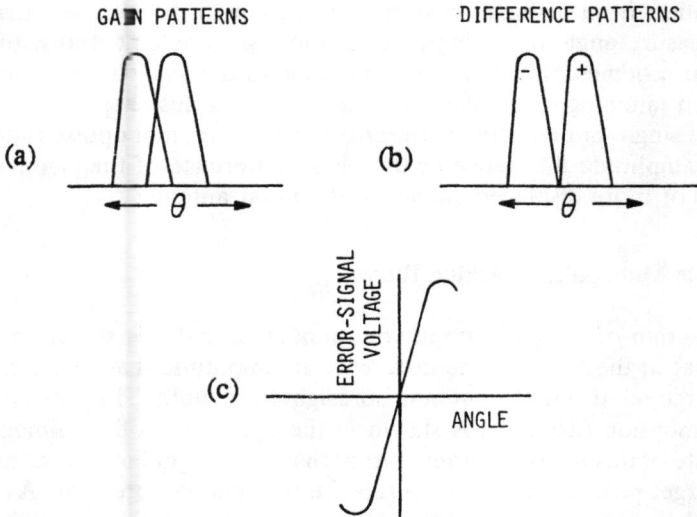

Figure 6.9 Amplitude monopulse error signal.

6.3.3 Phase Monopulse Tracking Radar

The block diagram for a *phase monopulse* radar is not unlike that for an *amplitude monopulse* radar: the primary distinction is that the Δ signal is due to the difference in the TOA rather than amplitude at two or more separated elements in the radar antenna. The result of this process is as shown in Figure 6.10. In the figure, we can see that if the signal source is away from the line perpendicular to the line joining the two elements, the phase of arrival at one element is different from that at the other element. Only when the signal source is exactly on the perpendicular line is the TOA of the signal at both elements the same. The TOA difference can be determined by measuring the phase difference between the two received signals. To avoid ambiguities in the angle measurement, the difference in phase of arrival must be less than the carrier wavelength; the extent of the ambiguities is dependent on the separation between the antenna elements. Theoretically, the phase measurement can be accomplished on each individual pulse at the TOA of the signals at the antenna elements.

As in the amplitude monopulse radar, an error signal is also generated in the phase monopulse radar. In this case, however, the error is a function of the phase difference, which, for a tracking radar, is true for targets having a displacement from boresight of less than a beamwidth. The angle tracking loop uses this error signal to adjust the pointing angle of the radar antenna beam so that a zero error results, which is when the boresight of the antenna is on the target.

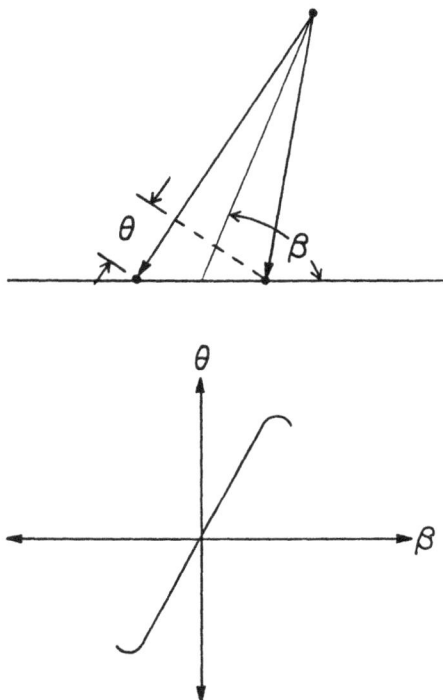

Figure 6.10 Phase monopulse error signal.

Because the monopulse radar can theoretically determine the angular position of a detected signal source on a single pulse, amplitude modulation of an ECM signal has little effect on these radars. Furthermore, it is not possible to determine at the ECM system whether the tracking radar is a LORO or monopulse radar because in either case there is no deliberate nutation of the radar transmitting antenna from which angle tracking error signals can be detected.

Because a monopulse radar can theoretically measure the AGA of intercepted pulses on a single pulse and is less vulnerable to amplitude modulation jamming techniques, ECM designers are required to develop new techniques of counter-measures against these radars. ECM techniques which were developed to jam sequential lobing radars are completely ineffective against monopulse radars; the techniques of inverse gain and swept square wave are in this category.

6.3.4   Cross-Polarization ECM

6.3.4.1   *Amplitude Monopulse Radar ECM*

Figure 6.11 shows one potential vulnerability of monopulse radars which can be exploited by ECM systems. As shown in the figure, the radar antenna patterns for a signal source whose polarization is orthogonal (90°) to the designed polarization of the radar antenna, distort the criterion used by amplitude monopulse radars. As shown, the nulls of the cross-polarized patterns of the two antenna beams occur where the properly polarized patterns are a maximum. The positions of the maxima of the two beams are interchanged in the cross-polarized patterns. This results in an error signal pattern which is 180° out of phase with the error signal established with the properly polarized signals. The net result is that for the same position in angle relative to boresight, the error signal for the cross-polarized signal is in the opposite direction from that required for proper boresighting, which would be provided with a properly polarized signal. This is illustrated in Figure 6.12.

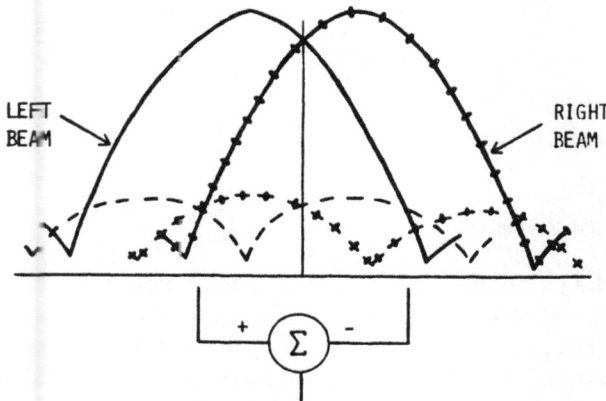

Figure 6.11 Effect of antenna polarization.

Therefore, in order to disrupt the operation of the angle tracking loop in a monopulse radar, an ECM system can measure the polarization of the intercepted radar signal and transmit signals which are orthogonal in polarization to that received. It is imperative that the ECM signal be orthogonal to within 1° of the radar receiving antenna. If not, the ECM signal will behave more like an augmenting signal than a jamming signal.

A problem associated with this technique is that the cross-polarized signal arrives at the radar antenna at its unpreferred polarization, at which the sensitivity

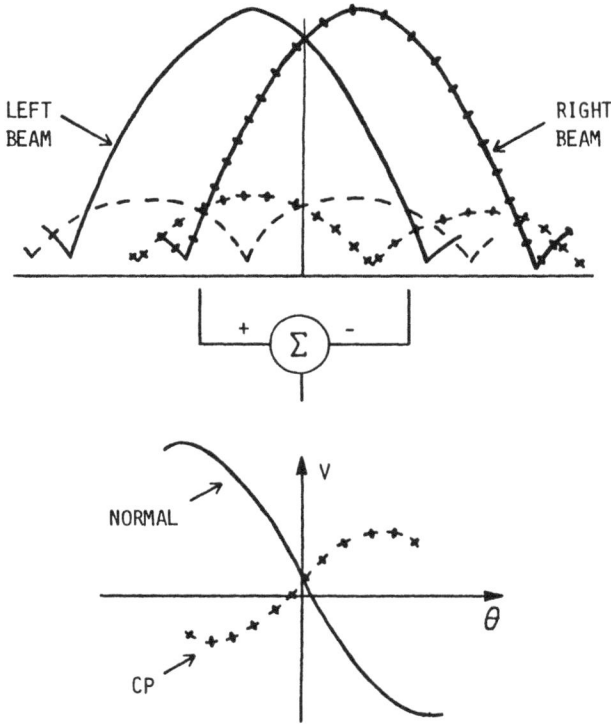

Figure 6.12 Amplitude monopulse error curves.

of the radar can be as much as 20 dB less than it is to signals arriving at the proper polarization. This places a formidable J/S requirement on the ECM system to compensate for this loss in sensitivity; typically, the required J/S is on the order of 30 to 40 dB, unless the RGPO or VGPO technique is used to support this angle jamming technique.

An extremely serious problem associated with the cross-polarization technique is that the orthogonality on the ECM transmitted signal is based on the measurement of the intercepted radar signal. Unfortunately, the intercepted signal is from the radar transmitting antenna, which may or may not be at the same polarization as the radar receiving antenna. If the radar receiving antenna is at least 1° different in polarization from the transmitting antenna, the cross-polarization technique is ineffective; the ECM system may operate as an augmenter rather than a jammer when this condition exists. There is no inherent reason that the radar receiving antenna must be designed to be at the same polarization as the transmitting antenna, because the radar reflections from typical targets are randomly distributed at all polarizations, and not necessarily at the polarization of the radar transmitting or receiving antenna.

6.3.4.2 Phase Monopulse Radar ECM

The cross-polarization mechanization used against amplitude monopulse radar is equally applicable against phase monopulse radars. Just as the error signal is inverted for the amplitude monopulse radar (Figure 6.12), a distortion in the error signal in a phase monopulse radar also results because of a 180° phase shift in the signal at one of the two paired elements in the radar antenna.

Figure 6.13 illustrates why this phase shift occurs. With a properly polarized signal source for the radar receiving antenna, the wavefront has a constant phase across the main beam of the antenna, as well as across each of the sidelobes in the antenna pattern. This is true except at the nulls in the antenna pattern. As we see in the figure, the phase front changes polarity from one sidelobe to the next. In a properly polarized signal, this polarity is correct for all of the main beam over which the tracking loop is designed to operate.

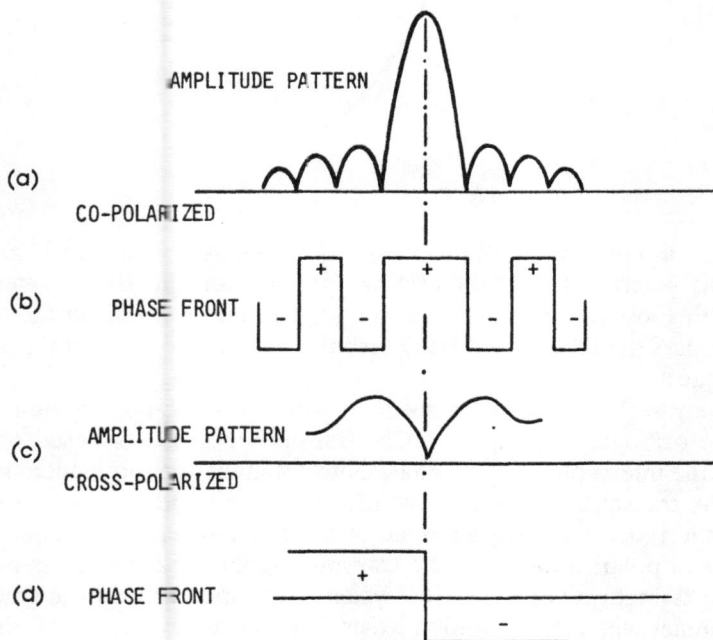

Figure 6.13 Wavefront distortion.

If, however, the signal source is cross-polarized, as indicated before, the null in the receiving antenna pattern is located on the expected boresight of the antenna. Therefore, as shown in Figure 6.13, the phase front shifts polarity immediately at the boresight of the antenna, which indicates a 180° phase shift in the signal TOA measurements and a distortion in the angle tracking loop error signal pattern.

This is the reason why the cross-polarization technique is often referred to as the technique which presents a warped wavefront to the radar. As seen in Figure 6.13(b), the phase front is constant over all of the main beam, whereas in Figure 6.13(d) the phase front experiences a 180° phase shift immediately at the boresight position of the antenna.

6.3.5 Cross-Eye

Another ECM method developed to counter monopulse tracking radars is the cross-eye technique shown in Figure 6.14. The intercepted radar signal is received at two ECM antennas at relatively distant points on the ECM vehicle. One of these signals is retransmitted (via a transmitting antenna) at the opposite end of the configuration without any additional delay. The other received signal is retransmitted through a transmitting antenna at the opposite end of the vehicle; this signal, however, is phase shifted in a manner that produces a 180° phase shift before retransmission. The net result is that the two signals arrive at the radar antenna at exactly the same time, but one is shifted in phase 180° relative to the other. This produces a null at the antenna when it is boresighted on the ECM vehicle, and, as shown with the cross-polarization patterns, a warping of the phase front occurs immediately at the design boresight of the antenna.

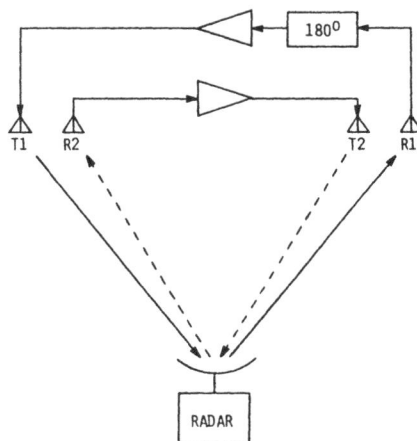

Figure 6.14 Cross-eye jamming.

The cross-eye technique is extremely sensitive to the amount of separation between antennas, and is usually used in relatively large vehicles where a large antenna separation is practical. This separation is measured on the perpendicular to the LOS between the radar and the vehicle. Therefore, a vehicle having an antenna separation parallel to the radar LOS poses a zero separation for this technique. The smaller is this separation, the larger is the J/S requirement for effective countermeasures.

6.4 J/S REQUIREMENT FOR THE MONOPULSE TECHNIQUES

Because the cross-polarization and cross-eye techniques depend on signals transmitted with distorted wavefronts relative to the radar antenna, they present nulls of the signal, which can be as much as 20 to 30 dB below that of the properly polarized true target signal. The ECM transmitter must therefore compensate for this loss in ERP when the techniques are designed to compete with the true target. However, if the RGPO or VGPO techniques are used to capture the range or velocity gates and move the tracking away from the true target position, only the 3 dB J/S for the gate stealing techniques is required. Having captured the tracking gates, the angle deception techniques enjoy an infinite J/S ratio. Because of the very high J/S requirement for the monopulse ECM techniques, capture of either the range or velocity gates is almost always required with the cross-polarization or cross-eye techniques.

6.5 OTHER ECM TECHNIQUES

6.5.1 Introduction

In addition to the techniques discussed previously, other effective ECM techniques have been devised to disrupt the operation of the angle tracking loops of tracking radars. These include the data rate reduction and ground bounce techniques. Several techniques have also been developed to exploit vulnerabilities in the logic circuits of these radars. Because tracking radars rely extensively on logic circuits based on fixed algorithms, they may be vulnerable to modulations which confuse or deceive these circuits. However, such ECM techniques are very fragile, because the algorithms used in the radar can be readily changed to counter them.

6.5.2 Data Rate Reduction—Sequential Lobing Radars

Sequential lobing radars require a predetermined number of pulses to extract the proper amplitude modulation used by the angle tracking loop to develop an error signal when the antenna boresight is off the target position. This is true whether the radar is actively scanning or is operating with a LORO antenna.

In Figure 6.15(a), the dotted line represents the amplitude modulation if all of the pulses are received in the angle tracking loop. If, however, only a few pulses are passed into the tracking loop, there may not be an adequate amount of modulation present with which to develop an error signal. This technique is known as the *data rate reduction* technique, and, in some applications, as the *count-down* technique.

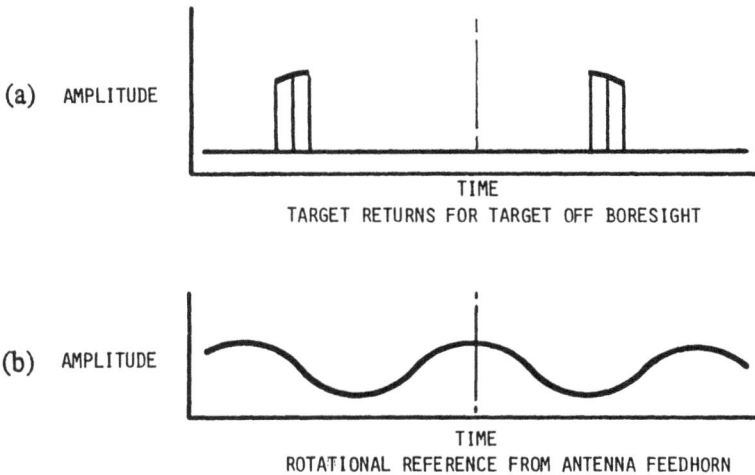

(a)   AMPLITUDE

TIME
TARGET RETURNS FOR TARGET OFF BORESIGHT

(b)   AMPLITUDE

TIME
ROTATIONAL REFERENCE FROM ANTENNA FEEDHORN

Figure 6.15 Data rate reduction.

The count-down nomenclature is derived from one of the methods of mechanizing the technique. The ECM system initially transmits a pulse for each intercepted radar pulse. During this time, the RGPO or the VGPO technique is applied to move the radar tracking gate (either the range gate or the velocity gate), away from the true target position to the hold-out false target position. When the radar tracking gate is positioned on the hold-out target, it passes only the ECM signal to the angle tracking loop, without interference from the true target. When this occurs, the number of pulses transmitted by the ECM system into the angle tracking loop is progressively reduced to the point where insufficient data is available to properly extract the amplitude modulation needed to develop the tracking error signals.

It is important that the data rate not be reduced so far that the number of pulses do not satisfy the radar requirements for target acceptance by the detection circuits. Tests have shown that the data rate can be reduced to about 10% of the radar duty cycle and still satisfy the detection requirements of the radar. With this low level of data rate, the angle tracking loop may not be able to develop an error signal. Indeed, the loop tends to operate as if no error exists, no matter how far off boresight the target is located. With this condition, the angle tracking loop coasts at the most recent proper error signal and, in most cases, will drift off the target position because the error signal produced by the data rate reduction technique implies good tracking by the loop.

Another effect which can occur in some radars with the data rate reduction technique is that the low data rate can cause the limiting of the signals in the radar receiver amplifier circuits. This is because the low data rate indicates a low average signal level in the AGC circuit, which then adjusts the gain of the receiver amplifiers to the point at which the pulses still present are operated at the saturation level of the amplifiers. This condition strips off the amplitude modulation due to the radar antenna nutation. The net result is that modulation on the pulses does not exist, even though the data rate may be adequate for the radar to extract such modulation if present.

6.5.3 Data Rate Reduction—Monopulse Radars

Although monopulse radars theoretically can measure the AOA on a single pulse, angle tracking loops used in these radars are still required to obtain an adequate data rate for proper angle tracking of the target. A high data rate is required to integrate signals out of the noise levels better, to provide a higher S/N in the loop. Furthermore, the number of pulses available in the loop has a significant impact on the tracking characteristics of the loop (e.g., tracking errors, tracking accuracy and response time). In addition, the AGC loop is generally designed on the basis of the reception of a predetermined number of pulses; a low data rate can severely affect the operation of this function. Therefore, the data rate reduction has been shown to be equally effective against monopulse tracking radars, even though these radars are reputed to be single-pulse angle measuring systems.

The data rate reduction technique is therefore very powerful because it can be used equally well against any tracking radar (conventional pulse, FMCW, or pulsed doppler) using either sequential scan or monopulse angle measurement. With the technique, it is not necessary to determine what type of radar is in the threat environment, because the mechanization is the same for all types. Furthermore, the mechanization of the technique is extremely simple, because it only requires amplitude modulation of the ECM transmitted signal. However, the technique must be accompanied with the RGPO or the VGPO technique to be effective, unless very high J/S levels (>20 dB) are used.

6.5.4 Ground Bounce

Another technique developed to counter radars which track in angle is known as the *ground bounce* technique (also called the terrain bounce technique). As shown in Figure 6.16, to use this technique, an ECM system located on the vehicle being protected directs its transmitter energy toward the surface of the earth in a manner that reflects the energy from the surface toward an interceptor. In this technique, the signal, which is unintentionally directed toward the interceptor from the ECM system, must not overwhelm the reflected signal so that the radar aboard the interceptor tracks that signal rather than the surface reflected signal. In addition, the surface reflected signal must be at a greater level than the radar echo signal from the vehicle being protected or the radar will track that signal because its TOA precedes that of the reflected signal (due to its shorter path of travel).

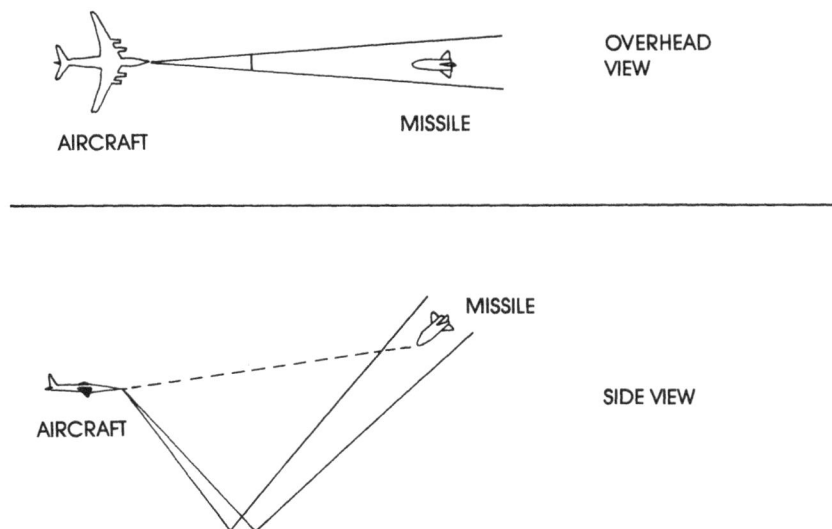

Figure 6.16 Ground bounce.

Assuming that the direct ECM signal and the target echo signal are at the required low levels, the interceptor radar will track the surface reflected signal and move toward the imaginary target, as shown in Figure 6.16. Because of the reflection characteristics of the surface, we can expect that the radar will track the elevation of the image target; because the surface reflection is diffuse in all other directions, little, if any, effect can be expected in azimuth tracking by the radar.

The ground bounce technique is primarily useful against missiles attacking low altitude vehicles, because these missiles are designed to approach low altitude targets from a higher altitude, and will be moving towards the surface anyway. Also, a low flying vehicle has a better chance of reflecting large power levels from the surface while minimizing the power transmitted in the direction of the attacker.

6.5.5 Expendable Electronic Decoy

One of the most powerful techniques available to designers for the protection of penetrating vehicles against radar defense systems is the use of expendable decoys. This device is applicable to surface or airborne vehicles. The effect on a victim radar is achieved by the remote location of the decoy from the protected vehicle. Although the effect on a victim radar is generally in all location parameters, the most significant is in the radar's target angle measurement.

Although passive decoys such as corner reflectors and chaff have been developed, the electronic decoy provides more flexibility because its radiated signal can be modulated to produce more realistic or confusing target parameters to the victim radar.

In the use of the decoys against tracking radars, the decoy's separation from the vehicle that it is designed to protect must be within the tracking capabilities of the victim radar. For example, if an acceleration of 3 g already exists between the vehicle and the attacking interceptor, any decoy separation which increases that acceleration (as viewed by the tracking loop) may cause the radar to remain locked onto the vehicle being protected. The characteristics of all radar tracking loops must be honored so that the decoy can be effective.

The significant advantage of the decoy is that it is effective against both sequential and monopulse angle tracking. The disadvantage is that the decoys are usually designed to be expendable, which dictates that the decoys be judiciously used. Especially important in airborne applications where a limited supply of such decoys is available is that they be used only when a terminal attack has been determined to be in progress.

6.6 TRACK-WHILE-SCAN (TWS) RADARS

6.6.1 Introduction

For accurately measuring and tracking target position in range as well as angle, in addition to narrow antenna beamwidths and pulsewidths, a tracking radar requires a high data rate of target returns. As a result, these radars are usually designed to lock onto their targets and dwell on the target position for the duration of the engagement. This is known as spot-lighting of the target because a pencil beam antenna is used to illuminate the target in this mode of operation. However, when the radar is required to track two or more targets simultaneously, it is done with some compromise to the tracking accuracy of each target, but not to such a degree that the effectiveness of the radar is seriously degraded.

A typical requirement for multitarget tracking is when the target tracking radar is also responsible for guiding an interceptor to the target. In this case, the target tracking radar must also maintain track of the interceptor at least in its initial and midcourse flight, and at the same time maintain accurate location of the intended target. The radar need not maintain track on interceptors that possess an autonomous midcourse or terminal guidance capability.

Antennas on older TWS systems are mechanically scanned and cannot be switched easily or rapidly between targets of interest. In this case, time is lost in scanning between the targets because of the mechanical inertia of the scanning systems. The antenna scan cycle is usually fixed in format, and the antenna beam is scanned at a relatively rapid rate to produce updated data on each target with as little delay as possible. The angular region scanned by the antenna is deliberately kept low (less than 60°) to maintain an adequate data rate on each of the targets.

With the advent of electronically steerable antennas, the multiple target tracking requirement is readily fulfilled by switching the antenna beam at very rapid rates (within microseconds) between all targets of interest. This results in the radar sharing the total possible dwell time between each of the targets, and thus yielding a compromise in the measurement accuracy of the position parameters of each target. Because a minimum amount of time is lost in switching between targets (as compared to the sequential scanning radars), the effect of time-sharing on the data rate is considerably smaller for the electronically steerable antenna radar.

6.6.2 Sequential Scanning TWS Radars

The operation of a typical sequential scanning TWS radar is shown in Figure 6.17. These radars are most often designed with measurement accuracy in only one dimension, in order to minimize the time away from the target position and increase the data rate in at least the one dimension. Because the azimuth dimension is more useful in the defense problem, a radar antenna fan beam, which is narrow only in that dimension, is used. As the antenna scans across its field of view, the power levels of echo signals are detected, as shown in Figure 6.17(b).

The display is much like that of the echo returns discussed in the chapter on search radar detection. As for the search radar, the echo pattern should closely approximate the radar antenna pattern (one-way or two-way) depending on the design of the radar.

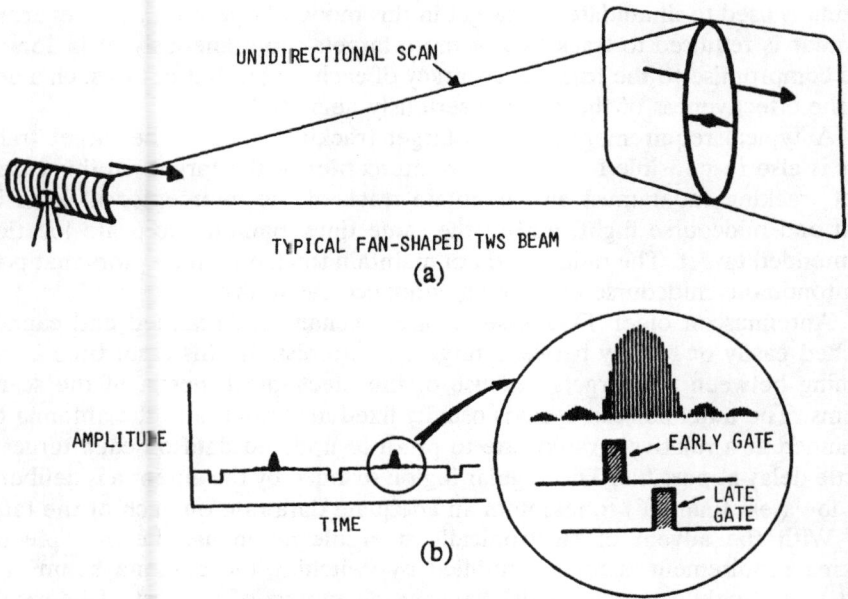

UNIDIRECTIONAL SCAN

TYPICAL FAN-SHAPED TWS BEAM

(a)

AMPLITUDE

TIME

(b)

EARLY GATE

LATE GATE

Figure 6.17 Track-while-scan radar.

Search radars are only required to measure the approximate main beam location, but *search-while-track* radars are required to measure where within the main beam the signal source is located (as is true for purely tracking radars). As shown in Figure 6.17(b), the radar uses what are referred to as early and late gates to locate the centroid of the main beam. In much the same manner as discussed with range tracking gates, the angle tracking loop determines a more accurate angular location of the target by forcing the loop to centroid on the energy contained within the two "windows." The argument is that the radar antenna is boresighted on the target when the energy in both gates is equal.

As shown, the return is from a single target at a particular range position. A target at another range position but the same angle can be differentiated from this target because of the difference in TOA of the signals; in that case, a similar pulse pattern would occur for the other target in the appropriate range gate of the radar. The radar range gates are used to determine which of the targets are processed by the angle track loop at any one time.

At another angle, even if at the same range, a target or the launched missile can be differentiated from another target because of its azimuth position. If the target is sufficiently removed in angle from another target, its signal return pattern may be isolated from the other target, even if range of each to the radar is the same. The target's echo pattern would appear at the appropriate angle in the display.

The TWS radar can maintain fairly accurate angle tracking on two or more targets. Although the accuracy of measurement is not as good as that achieved with a spot-lighting radar, it is an acceptable compromise to achieve multiple target tracking capability with a single radar. The more targets which are tracked simultaneously by a single TWS radar, the more serious the degradation in performance against each of the targets.

6.6.2.1 ECM versus Sequential Scanning TWS Radars

Because a TWS radar operates much like a search radar with a continuous scan over its FOV, the techniques of jamming the range measurement function (as suggested for the search radars), apply equally well to this radar; these include the noise and multiple target techniques during the mainbeam period over the target position, as well as during radar antenna sidelobe illumination of the target.

However, TWS radars are required to measure their target positions much more accurately than early warning and acquisition search radars, and thus employ more sophisticated techniques of angle measurement than search radars. To degrade the performance of the TWS radars, ECM techniques have been devised to disturb their angle tracking function. Although the techniques developed generate relatively small errors in the target angle position measurement made by the TWS radars, the errors are sufficient to cause serious disturbances in the interceptor guidance function of the radar, especially in the guidance of unmanned interceptors.

Figure 6.18 shows an *inverse power programming* technique which is used during the main beam illumination of the vehicle carrying the ECM system. In a similar fashion as that used with a search radar against the angle measurement function, the ECM transmitter power output is programmed to vary in an inverse manner to that detected from the scanning radar as it scans across the vehicle position. If properly programmed, the modulation due to the scanning antenna pattern (as detected in the radar angle tracking loop) is removed; the tracking loop centroid process, which is required in this radar, is thus foiled. As seen in Figure 6-18(c) the power appears to be equally distributed over the complete range of the antenna pattern. When this occurs, equal power is measured in each of the two radar gates independent of the boresight position of the antenna.

In actuality, the amplitude level of the jamming signal is not as constant as visualized in Figure 6.18(c). Nevertheless, sufficient disturbances are produced in the centroid process so that adequate missile miss distances can be achieved. The advantage of this technique is that inverse power program is primarily designed to operate in the radar antenna's main beam, which greatly relaxes the ERP requirements (as compared to operation into the radar antenna sidelobes).

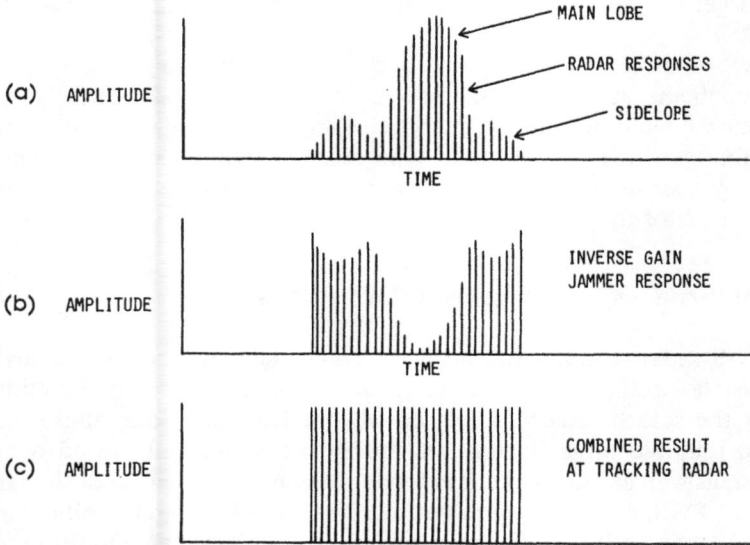

Figure 6.18 Inverse gain *versus* TWS radar.

A variation of this technique is shown in Figure 6.19. As indicated, the ECM transmitter operates on only one side of the detected mainbeam pattern. As shown, the radar angle tracking gates track with an offset from the true target angle. A modification to this technique is to alternate the ECM transmission from one side of the antenna pattern to the other at a relatively low frequency (on the order of one hertz). This "wobbling" of the target position has been shown to disrupt the guidance instructions from the radar to the missile.

6.6.3 TWS Radars Using Electronically Steerable Antennas

As previously defined, a TWS radar is one which maintains a search mode of operation with intermittent dwells on targets of interest. The dwell time on target is made sufficiently long to produce reliable and accurate measurements on the target position in angle and range. This capability has become more practical in recent years because of the development of the *electronically steerable antenna* (ESA).

Because the ESA beam can be randomly and rapidly switched (within a fraction of a microsecond) in its pointing angle, it is now possible to use a non-periodic scanning program with very flexible interruptions of the scanning cycle. These interruptions can be programmed to switch the antenna beam back and

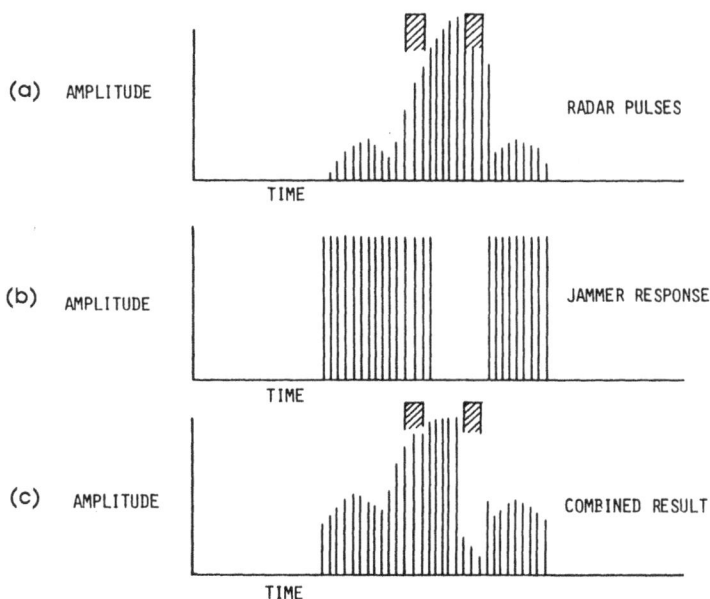

Figure 6.19 Sidelobe jamming of TWS radar.

forth to one or more angular positions where targets have been previously detected. The dwell times at these positions can be made sufficiently long to ensure accurate measurement of the target parameters. The inclusion of high speed computers and signal processors has made these types of radars more the norm rather than the exception.

Indeed, modern tracking radars are being developed which use electronically steerable antennas in their purely tracking mode. These radars are capable of simultaneously tracking several targets of interest by programming the antenna to switch to only the angular positions at which targets have previously been detected.

The angle measurement is most often made during the dwell times on target with the use of the monopulse technique, either amplitude or phase. As indicated previously, to counter this radar, an ECM system must resort to one or more of the techniques described as being effective against this type of radar. These techniques include cross-polarization of the ECM transmitted signal or the data rate reduction technique. In either case, one or the other of the radar gate capture techniques (RGPO or VGPO) must first be applied.

6.6.4 Downlink ECM

TWS radars are generally used when the radar tracking a target is also required to track the position of a missile launched toward the target which is to be guided in at least one of the phases of interception (e.g., at launch, in midcourse, or in its terminal phase).

In the case of missile guidance with the use of TWS radars, the radar is required to maintain both range and angle track on the missile, as well as on its target. Because the radar cross-sectional area of such missiles is relatively small (which may result in unreliable detection and tracking of the missile), a signal augmenting transponder is usually installed on the missile. The objective of the transponder is to return a signal which is much greater in power than that reflected off the missile, and which is reliably detected and tracked by the radar. The transponder signal is designed to emulate the reflected signal, and contains the same modulation in time, amplitude and frequency as the reflected signal. Measurement of these modulations by the radar will result in the true location of the missile.

If an ECM system aboard the protected vehicle can transmit a signal at the carrier frequency of the transponder (usually the same as the radar) and of sufficient power so that it overrides the transponder signal as it enters the radar receiver, the radar missile angle tracking loop can be made to track the target as the position of the missile. When this occurs, the radar sends guidance signals to the missile, which are based on the position of its target rather than on the current position of the missile.

Figure 6.20 depicts a typical ECM transmission used to counter the radar's missile tracking loop. Because the radar is designed to measure both the range and angle of the missile, the ECM program must include signals which arrive at the radar receiver at the same time as the transponder signal, which will very likely be at about the same time as the low power reflected signal from the missile. Because it is not possible at the ECM system to determine when the transponder signal enters the radar receiver, a series of pulses are transmitted (as shown in the figure) with the expectation that one enters the receiver at about the same time as the transponder signal, or at a time when one of the pulses may capture the attention of the missile tracking loop. Once this occurs, it is likely that the radar will accept and track on this signal in both range and in angle.

The launched missile being within the same beamwidth as the radar target is neither necessary, nor likely, until the final phase of its flight. It is, however, necessary that the ERP of the ECM system located aboard the protected vehicle be adequate to overcome the fact that the ECM transmitted signal used to counter the transponder signal may initially be generated into the sidelobes of the radar antenna beam when it is pointed at the missile. Until the radar's missile angle tracking loop is captured by the ECM signal, the ECM is effectively transmitting into the sidelobes of the radar antenna.

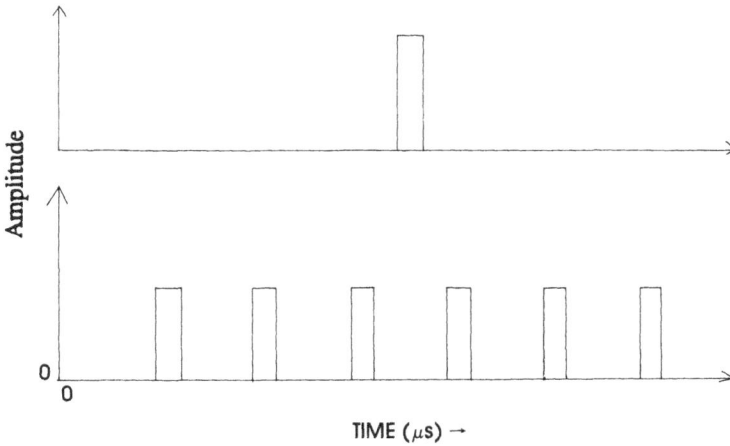

Figure 6.20 Downlink jamming.

The ERP required for this type of ECM system is as follows:

$$P_{rm} = \frac{P_m G_m}{4\pi r_m^2} \frac{G_{ML} \lambda^2}{4\pi} = S_m$$

where

$$P_m = \text{missile transmitter power}$$

$$G_m = \text{missile transmitter antenna gain}$$

$$r_m = \text{radar range to the missile}$$

$$P_{rj} = \frac{P_j G_j}{4\pi r_j^2} \frac{G_{SL} \lambda^2}{4\pi} = J_m$$

$$\frac{J_m}{S_m} = \frac{P_j G_j}{P_m G_m} \frac{G_{SL}}{G_{ML}} \left(\frac{r_m}{r_j}\right)^2 \frac{1}{(4\pi)^2}$$

$$P_j G_j = (4\pi)^2 P_m G_m \frac{G_{ML}}{G_{SL}} \left(\frac{r_j}{r_m}\right)^2 \frac{J}{S} \tag{6.1}$$

As indicated in the equation, the amount of ERP is dependent on the radar antenna mainbeam-to-sidelobe ratio, because it is most likely that the missile (at least at the beginning of the jamming), may not be within the same beamwidth as the vehicle carrying the jammer. Also evident in the equation is the fact that the amount of ERP required decreases as the missile approaches the target vehicle since r_m, the radar range to the missile, is approaching r_j, the radar range to the target, in the terminal phase of the intercept.

6.7 SUMMARY

Of all the target position measurements made by radars, the most significant for successful interdiction of vehicles penetrating enemy territories is the angle measurement made by the tracking radars associated with defense systems. Even if this was the only measurement possible in the engagement, interdiction is still possible, albeit with somewhat reduced performance. Achieving any reasonable degree of success in interception of penetrating vehicles is hardly possible if this parameter is not available for proper launching or guidance of the interceptors, either manned or unmanned.

Furthermore, because of the required angle accuracy with the tracking radars, the measurements are almost always accomplished with the use of automatic processing circuits. We have seen that radars operating primarily on the basis of signal processing techniques are more vulnerable to external interference than those radars which are manually operated or with a manual override of the electronic processing.

This chapter described the various techniques used by tracking radars to measure the position angle of its targets. In addition, the vulnerabilities of the radar techniques were identified with ECM techniques described which exploit these vulnerabilities. Because of the importance of this target position measurement, much effort has been expended by radar designers to preserve the capability by designing antijamming techniques that counter any newly developed ECM techniques.

The monopulse radar angle measurement techniques have presented a formidable challenge to ECM system designers. Although ECM techniques to jam that type of radar have been described, we also pointed out that these were suggested as possible solutions in spite of significant problems associated with their use. The cross-polarization technique, for example, requires a formidable amount of ERP in the ECM transmitter; moreover, if the polarization measured is not that of the radar receiving antenna, no amount of ERP will adversely affect the angle measurement capability of the radar.

Chapter 7
Other Radars

7.1 INTRODUCTION

In this chapter, we discuss radars that are not necessarily included in the class of search or track radars, but use techniques of measurement and tracking of their target's position similar to the ones used in those radars. In many cases, the ECM techniques used against the search and track radars are equally applicable to these radars, but they require special consideration because of the peculiarities of their operation.

The radars we discuss in this chapter are:

- Interceptor Guidance Radars
- Proximity Fuzes
- Synthetic Aperture Radars

Although there are other applications of radar technology, such as *terrain following terrain avoidance* (TF/TA) radars and radar altimeters, unlike the radars discussed in this book, these are not a direct threat to the survivability of penetrating vehicles. Although the capabilities of these secondary radars are crucial to mission success, little attention has been paid to developing techniques which interfere directly with their operation.

7.1.1 Interceptor Guidance Radars

Radars installed aboard intercepting vehicles are generally regarded as *terminal guidance radars*. The target position measurements during the final phase of the interception are performed aboard the interceptor, whether it is manned or unmanned. Two types of operation can be used with terminal guidance radars, active or semiactive. In an active radar, the transmitter, the receiver and all required processing circuits are located aboard the interceptor; the active radar requires little or no cooperation with the launching or the midcourse guidance radar once the interceptor radar detects and locks onto its target. Active terminal guidance radars are most often used with manned interceptors and with some large missiles.

In a semiactive radar, only the radar receiver and the processing circuits are installed aboard the interceptor. The transmitter of a semiactive radar system is generally located on the ground or another vehicle and, is usually the same transmitter used to track the target for launching and the midcourse guidance of the interceptor.

The fire control radar can "launch and forget" an interceptor if the interceptor operates with an active radar which has locked onto its target prior to launch; not so with interceptors using semiactive radars. Because the semiactive radars require radar illumination of the target, the fire control radar (or an associated radar), must continually track the target position in order to provide proper target illumination for the semiactive radar.

Furthermore, the semiactive radar, in most cases, must be able to receive reference signals from the illuminating radar to be able to properly process the signals reflecting from the target and detected in the radar receiver. Semiactive radars are best used in expendable missiles, because the transmitter, which is generally the most expensive part of a radar system, is not lost with the missile.

7.1.1.1 ECM versus Interceptor Active Radars

All of the techniques discussed in the section on search radars are equally applicable against the acquisition mode of active radars. Even though the fire control radar is able to guide an interceptor into the vicinity of its target, the interceptor radar must still search over a limited volume of space before locking onto its designated target. The noise and multiple target ECM techniques used against search radars are applicable against this mode of the interceptor radar.

Techniques of gate capture (either range gate or velocity gate, depending on the radar) can be applied reliably to active or semiactive radar guidance systems. This is because the time of interception of the pulse at the ECM vehicle is simultaneous with the time of reflection of the illuminating signal from the vehicle. Furthermore, the doppler frequency of the signal detected in the ECM receiver is exactly the same as that reflected from the vehicle due to the illumination by the ground radar. Any offset in time or frequency placed on the ECM signal is a true offset as measured in the interceptor receiver.

When the active radar aboard the interceptor has locked onto the target, the radar tracks in angle and in range much like the tracking radars discussed in Chapters 5 and 6; and ECM techniques discussed in those chapters are equally effective against the interceptor's radar. The only difference in the ECM design is due to the variation in the dynamic characteristics of the engagement between two moving vehicles—the target and the interceptor.

7.1.1.2 ECM versus Interceptor Semiactive Radars

Techniques such as RGPO and VGPO are effective against the semiactive radar, because the illuminating signal which is reflected from the vehicle to the interceptor is also detectable at the protected vehicle. However, because there is no radiation from the missile using semiactive guidance, it is not possible to determine the type of angle tracking the semiactive radar is using at the protected vehicle. The angle tracking function in the semiactive radar must then be assumed to be with a LORO or monopulse antenna.

Because there is no radiation from the interceptor in the semiactive mode, techniques of angle deception which require measurement of the characteristics of the intercepted signal are not effective. These include jog detection as well as the cross-polarization and cross-eye techniques. However, the swept square wave (without jog detection) and data rate reduction techniques are effective, and the ones most often used.

A semiactive radar is vulnerable because the illuminating radar must continuously track the target to cause a target reflection that is detectable at the interceptor. An ECM system aboard the vehicle being protected can exploit this vulnerability by jamming the tracking radar associated with the illuminating transmitter. By driving this radar off its intended target using the techniques discussed in Chapters 5 and 6, insufficient illumination of the target will result. The resulting reflection may not be adequate for proper detection and processing by the semiactive radar.

7.1.1.3 Track-via-Missile Radar

In the track-via-missile mode, target data (as collected at the interceptor) is transmitted back to the launch site, where it is processed to determine the interceptor-to-target spatial relationship. This mode of radar operation is used to provide operator or surface radar control of the terminal guidance of an unmanned interceptor. During engagement, the operator at the launch site has a continuous display as to the position of the interceptor missile relative to the designated target. Manual or launch-site correction of the terminal guidance can readily be applied at the launching site when necessary.

The location of the signal processing at the launch site, rather than aboard the missile affects the range, doppler and angle tracking functions of this type of radar only secondarily. The radar at the launch site operates as if it were a standard tracking radar, except that it also includes measurement of the parameters of the missile relative to the target using data collected at the missile. The time-of-arrival and doppler frequency of the target signal as detected at the interceptor are relayed back to the launch site. In addition, the target angle modulation parameters, as measured at the interceptor, are also used at the launch site to determine the angular relationship between the target and the missile.

The ECM system is not required to be aware that the guidance is being performed via the missile. ECM operation against this missile as if it were a semiactive radar would have the same deleterious effect on the guidance system albeit via the missile to the launch site. In this case, however, manual override of the operation of the missile system must be a consideration in the ECM design.

7.1.1.4 Home-on-Jamming Radars

Terminal guidance radars generally include a home-on-jamming mode, during which the radar concedes that the true target reflection is no longer reliable for target position measurement due to the ECM interference present. In this case, the radar angle tracking loop switches to operation on the interfering signal, recognizing the fact that its primary sorting mechanism, TOA or doppler frequency is no longer a reliable measure of that parameter. This mode of operation is applicable only to vehicles which are designed for this form of interception profile.

Because the source of the ECM signal is the ECM transmitting antenna, more accurate measurement of the target angle may result when compared to the radar's measurement of the erratic target reflection. Therefore, an ECM system aboard the protected vehicle must possess adequate J/S and an angle jamming technique to confuse or deceive the angle tracking loop employed by the interceptor.

7.1.2 Missile Fuze

7.1.2.1 Introduction

The engagement with the radar defense system should not reach the point where an interceptor missile is approaching the vehicle protected by the ECM. We would expect that one or more of the ECM devices previously described had been sufficiently effective to prevent proper launching of the missile, or had induced enough errors in the target position measurement to abort proper interception. Nevertheless, for the sake of completeness in our coverage of the engagement process, we include an examination of the final possible phase of the engagement (i.e., the missile proximity fuze).

We will now describe the fuzing process which may be used in an intercepting missile, with possible methods of deception or jamming of this process. It is not the intent of this discussion to imply that any missiles, friendly or hostile, use the fuzing techniques discussed. It is only suggested as a potential mechanization technique which may be used to offer possible counters if they are used. Nevertheless, we can realistically assume that all missile proximity fuzes contain some form of the suggested technique, the primary difference being in the fuzing parameters used.

7.1.2.2 Missile Proximity Fuze

Figure 7.1 shows a concept which can be employed for proximity fuzing of an attacking missile. The antenna pattern shown is a cross-section of a complete pattern, which is a volume of revolution about the axis of missile travel. Generally this pattern is coincident with the "kill" pattern of the missile warhead. A radar transmitter and receiver are associated with this antenna. When a reflection from the target is detected to be within the antenna pattern, detonation of the warhead will result in a kill of the target.

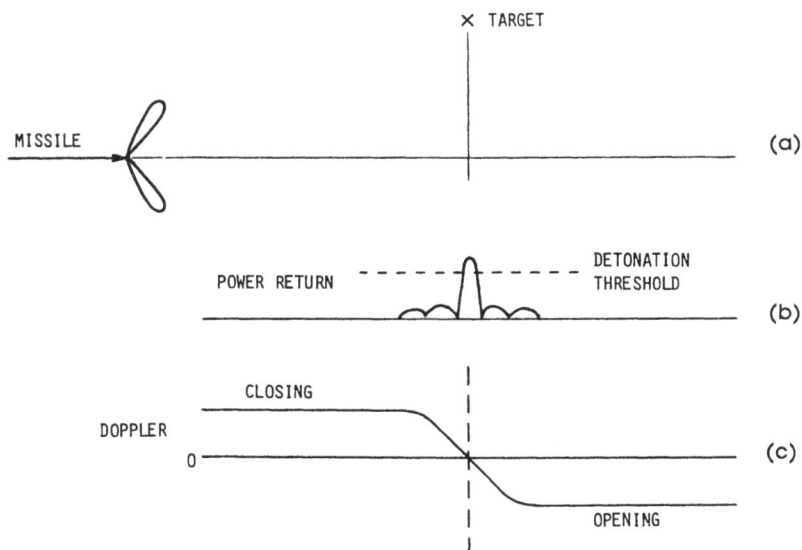

Figure 7.1 Missile Proximity Fuze.

As shown in Figure 7.1(b), a threshold is set in the fuze receiver, and, as the missile approaches the target position, the receiver in the missile is presented a signal return pattern that is in accordance with the missile antenna pattern. As shown, only when the target is within the main beam of the pattern does the signal exceed the threshold set in the missile receiver. Thus detonation takes place only when the target is within the antenna pattern, as is required. Because the target is also in the kill pattern, successful destruction of the target is expected.

For more reliable detonation, in addition to the amplitude threshold setting the transmitter system aboard the missile can be operated coherently as a doppler radar. As shown in Figure 7.1(c), with this type of operation, the missile can also detect the doppler frequency of the return signal due to the relative motion between the missile and its target.

At the beginning of this encounter, the doppler frequency (as detected by the fuze) is positive due to the closing nature of the engagement. As the missile passes the target, the doppler frequency passes through zero (the point of closest approach) and goes negative due to the opening nature of engagement beyond the target. The zero doppler frequency thus occurs at the point of closest approach to the target. The doppler frequency of the reflected signal can then serve as a reliable indicator of the missile's proximity to its target. This data can be used independently for warhead detonation, or be combined with amplitude threshold detection to provide a more reliable indication that the target is within the kill pattern of the warhead.

7.1.2.3 ECM versus Fuzing

A technique which may be used to deceive the type of missile fuze described above is shown in Figure 7.2, and is designed to detonate the fuze before it comes close to the target. As indicated, it is only necessary to retransmit a signal, either noise

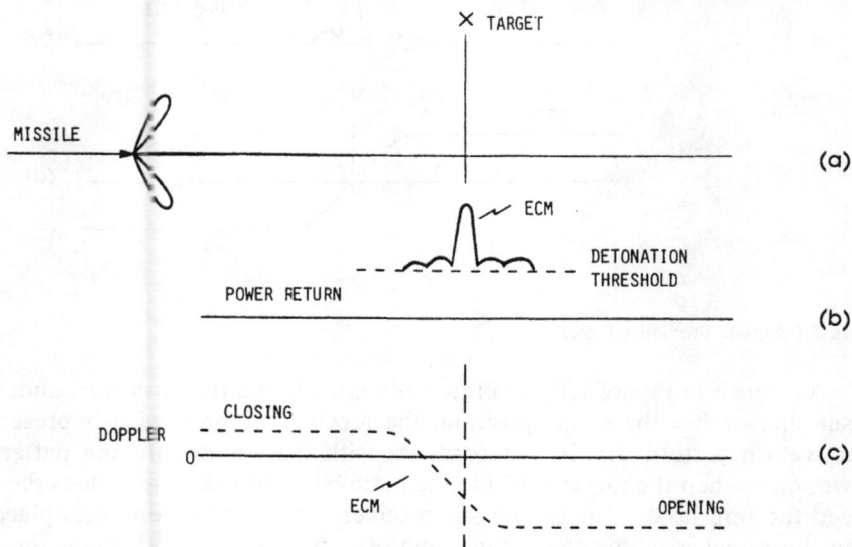

Figure 7.2 Proximity Fuze Jamming.

(which includes the fuze frequency) or a repeated signal at an amplitude which compensates for the lack of gain in the antenna sidelobes, when the missile is still distant from the target position.

At the same time, with a repeated signal, the doppler frequency can be shifted so that the zero crossing occurs some time before the missile reaches its target, resulting in premature detonation of the fuze. Shifting the doppler so that the zero crossing occurs after the missile has passed its target can be used to prevent detonation of the fuze. We note that, in the latter case, the missile fuze radar may yet be capable of detecting the true doppler from the target, resulting in a successful kill by the missile.

7.1.3 Synthetic Aperture Radars (SAR)

7.1.3.1 General

Synthetic aperture radar (SAR) was first introduced as a ground mapping radar because of its ability to resolve very small reflective areas. This was made possible with the use of very narrow effective beamwidths and high resolution in range. The narrow beamwidths were accomplished with the use of synthetically produced large antenna arrays. A critical requirement for these radars is that the reflector is nonmoving, because of the requirement for strict adherence to the proper phase relationship of all reflections during the required scan time of the synthetic aperture antenna. For ground mapping, these requirements were readily observed.

Ground forces strategists have determined that a more effective attack could be waged against the enemy if it were possible to measure the extent of the enemy forces, the types of vehicles being used, and the location and movement of the vehicles. They decided that this requirement could be met with the use of a variation of the ground mapping radar using a synthetic aperture antenna, especially because the vehicles of interest moved at a small enough relative velocity to satisfy the phase coherency requirement.

7.1.3.2 Description

To provide high resolution of targets in the range parameter, radars use very narrow pulsewidths (on the order of nanoseconds) or pulse compression techniques which result in even more narrow pulsewidths. The pulse compression technique was described in Section 2.12 as a technique used to produce high average transmitter power levels with relatively low peak power levels, as well as provide high range resolution with resultant short pulsewidths. By using a long coded pulse to maintain a relatively high average power transmission, the radar can decode the

172

target reflection to yield a very narrow pulse, and thus a high resolution in the range parameter. This technique has been applied effectively to radar tracking of low profile missiles and targets. Target resolution capabilities as low as a few meters have been realized with the pulse compression technique.

To provide high resolution in the target angle parameter, a very large antenna aperture is required because the resolution of reflectors is equal to

$$R \cdot B = R\lambda/D$$

where

R = range to the target;
B = antenna beamwidth;
λ = radar carrier wavelength;
D = aperture size.

For a target resolution of 100 m at a range of 100 km and a wavelength of 3 cm, the antenna aperture width must be at least 30 m wide, which is most often too large to be practical for the application.

As shown in Figure 7.3, a phased array antenna achieves its narrow beamwidth by coherently summing the signals which arrive at the distributed elements of the antenna array. Only a signal within the desired angle adds up coherently, yielding maximum gain in that direction. In this case, the summing of the signals is performed simultaneously, and, if the width of the antenna aperture is 30 m, a target resolution of 100 m at a range of 100 km can be achieved. As seen in the equation, target resolution depends on the range to the target.

Figure 7.3 shows that the phase of arrival at each element of the antenna array is dependent on the distance of the element from the center of the array. The array must be designed so that the phase difference at the outermost element is less than a quarter of a wavelength from the difference at the center of the array. As shown in the figure, the phase shift is a function of the distance of each element from the center of the array. The phase shift can be shown to be

$$d(x) \approx \frac{2\pi x^2}{\lambda R(0)} \tag{7.1}$$

where

$R(0)$ = nearest range to the target;
x = position of element relative to element position nearest to the target.

If a phase shifter is now placed in series with each element of the phased

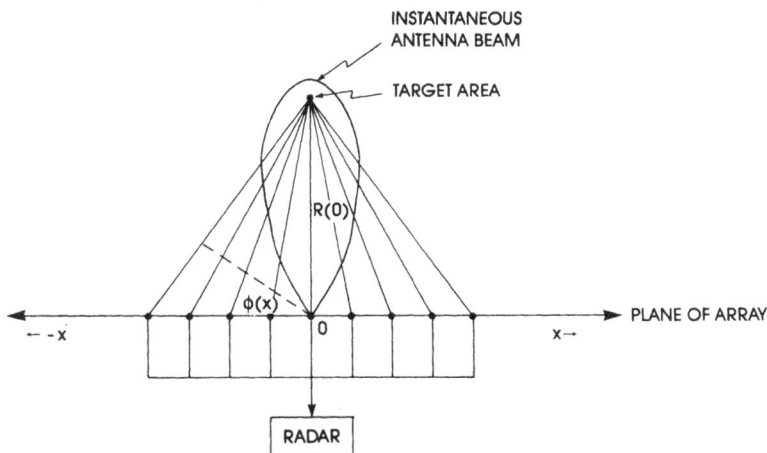

Figure 7.3 Unfocused antenna phased array.

array (as shown in Figure 7.4), it can be adjusted to compensate for that phase difference so that the signals from all elements of the array arrive in exact phase at the summing circuit of the processor. This adjustment in phase is referred to as focusing of the antenna. Because the phase adjustment is dependent on Equation (7.1), we can see that it is dependent on the true range to the target, $R(0)$. Thus, each set of focusing adjustments is for only one target range position at a time.

Figure 7.5 shows a method which can be used to achieve high resolution in angle, but with much smaller aperture sizes. As shown in the figure, only one of the elements of the antenna array is used at any one time. This element is moved along the line parallel to the face of the synthetically produced broad array. At each equivalent element spacing of the ultimate array, a pulse is transmitted. At each pulse transmission, the radar receiver detects, measures and stores in memory the phase of the target return. This process is continued until the antenna element reaches a point which is equivalent to the farthest element of the synthetically produced array. At this point, the radar processor sums the signals and, as with a real array, only those signals which arrive from the design angle add up coherently, providing maximum antenna gain in that direction. This process results in a wide antenna aperture which is produced synthetically in a sequential manner rather than on an instantaneous basis.

The beamwidth of the single element array must be broad enough that the signal, even though it may be outside the beamwidth of the synthetically produced beam, is within the single element beam. Furthermore, the phase of arrival of the signal from the edges of the synthetic antenna must not be larger than an eighth

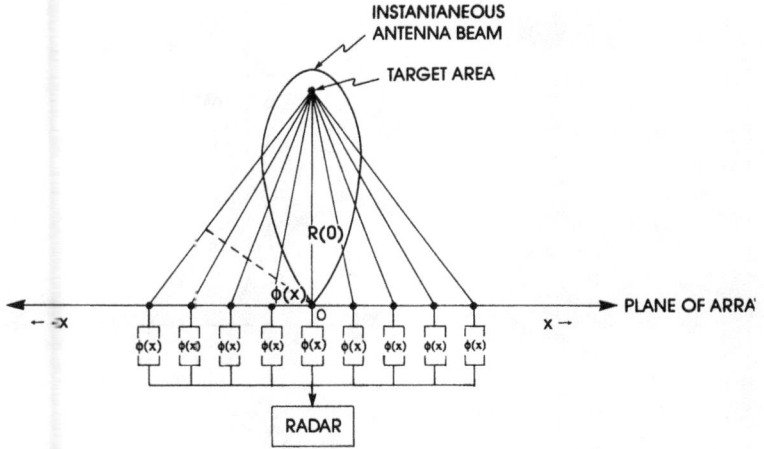

Figure 7.4 Focused antenna phased array.

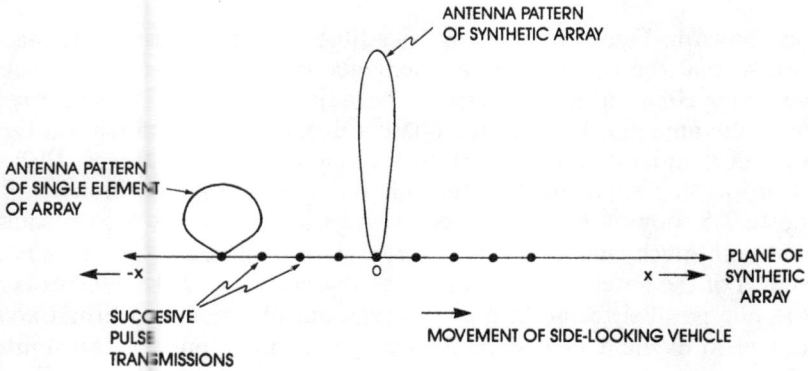

Figure 7.5 Synthetic aperture antenna.

of a wavelength from the phase at the center of the beam. Because of the two-way path for this array, the net result is a quarter of a wavelength difference from the outermost element to the element nearest the target position. This satisfies the criterion for coherent integration of the reflected signals.

Although the above description includes only one scan of the element array, this is generally a continuous process implying an infinitely long array. The radar processor, however, operates only on predetermined time intervals that are consistent with the resolution, and thus the required size of the array. With side-looking radars, the array element is most often positioned on an airborne vehicle so that its main beam is pointed orthogonally to the path of travel of the vehicle. The radar processor determines its rate of sampling by noting the speed of the vehicle and the radar pulse rate.

As with the real aperture size antenna, there is a difference in the TOA of the signal when the signal element is at different points of the array path. This is not different than is shown in Figure 7.4. This phase difference is extremely small; indeed, it must not vary more than an eighth of a wavelength during any one complete scan to provide the resolution required. If the radar processor corrects for this phase difference at each element position, the signal will add coherently at the processor only if it is at the position (in angle and range) indicated. This operation is referred to as focusing of the synthetic aperature antenna to a point in space.

Because the focused antenna restricts its detection to a point in space, its angular resolution is shown to be

Resolution = $D/2$

The process of focusing the antenna beam results in a resolution which is independent of the range to the target and is dependent only on the real width of the single element antenna. For example, focusing a synthetic aperture antenna using a real antenna with a 10 m aperture provides a resolution capability of 5 m. Although the focusing requirement is dependent on the radar's range of interest, the target resolution in angle is independent of range.

7.1.3.3 ECM versus Synthetic Aperture Radars

A significant vulnerability of SARs exists because the narrow beam (focused or unfocused) is not formed until a predetermined number of pulses are captured by the radar. This is unlike a real antenna where the resultant beam is formed instantaneously. At each element position in the SAR, the real (broad) beam is in effect, so that an interference signal, which may appear to be in the sidelobes of the synthetically produced beam, is still within the half-power beamwidth of the uncompressed antenna. Figure 7.6 illustrates the relationship between the instantaneous beamwidth of the single element antenna and the compressed antenna which results.

The net result is that, before the radar is able to produce the synthetic beam, the target signal and the interfering signal are within the same antenna beamwidth (the broad beam of the single-element antenna). All circuits in the radar receiver which occur before the signal processor are subjected to the J/S achieved on each pulse before compression of the antenna beam is achieved. We can expect that all such functions will be seriously degraded before the processor can produce the compressed beam.

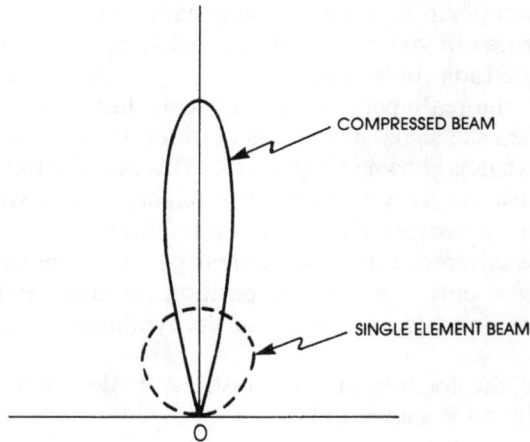

Figure 7.6 Compressed beam relationship.

A noise jamming signal at the location of the protected target can produce an interference signal in the front end of a SAR receiver which suffers no more than 3 dB in *J/S* because of its location at either end of the real antenna beam. Because the width of this beam is generally about 30 to 40 times that of the synthetically produced beam, the net result is that the radar will be able to resolve targets in angle at this location to no better than 30 to 40 times the desired resolution. In comparison, the same noise jammer against an antenna aperture (which produces the compressed beamwidth instantaneously), must also increase its transmitter power to overcome the loss of antenna gain in the sidelobes of the real antenna (on the order of 20 to 30 dB).

We must remember that this radar almost always operates with pulse compression to produce high resolution in the range parameter. The ERP of a noise jammer must be increased to overcome the loss (usually about 20 dB) inherent in the processing gain associated with pulse compression.

A continuous noise jammer (i.e., one which transmits noise over each complete pulse interval of the radar) will protect all targets contained within the beamwidth of the broadbeam real antenna, assuming the ECM system produces an adequate power level to overcome the loss due to pulse compression by the radar. The net result at the radar is a noise strobe which encompasses the complete beamwidth of the real antenna; as previously stated, this beamwidth is about 30 to 40 times that of the design beamwidth of the synthetically produced antenna.

Chapter 8
Resource Management in ECM Systems

8.1 INTRODUCTION

The science of electronic countermeasures can trace its roots to the simple broad-beam, broadband, noise jammer, which was an extremely effective system until radar designers developed techniques to counter that type of jamming. The advantages of such a jammer are numerous. It provides simultaneous capability against all radars within the beamwidth of the jamming antenna and thus requires no AOA measurement of the intercepted signals, nor steering of the ECM system transmitter antenna. The jammer operates effectively against all radars that have their operating carrier frequencies contained within the jamming frequency range, thus eliminating the need for interception and frequency measurement of the radar signals. Also, frequency set-on of the jammer transmitter is not required. Equally important is the fact that such a jammer is immune to radar advancements such as carrier frequency and PRF agility. Because interception and measurement of the incoming radar signals is not necessary, no receiver is required, thereby eliminating the need for interruption of the jamming for the purpose of receiver look-through.

8.1.1 Power Sharing

The broadbeam barrage type of jammer, however, suffers from an extreme waste of the limited available transmitter power. First, the frequency range of jamming is dictated by the lowest and highest expected carrier frequency of operation of all radars within the angular coverage of the jammer's transmitting antenna. Too often only a small percentage of the jamming frequency range is used by the radars of interest in any one engagement. The power dilution for any one victim radar is equal to the ratio of the detection bandwidth of the radar to the bandwidth of the jamming, independent of the number of victim radars in the environment.

Second, the jamming antenna beamwidth is designed to encompass the complete angular region within which all threat radars are expected to be located. More effective use of the radiated power would be achieved with an antenna which is pointed in the direction of the victim radar. For any one radar, therefore, the power dilution due to the space factor is equal to the ratio of the gain of the broadbeam jammer antenna to the gain of the directive antenna which could be used.

A third factor affecting optimum use of available jammer transmitter power is the time period of the jamming. When operating against search radars, the brute force jammer is operated even when the radar receiving antenna is directed away from the jamming vehicle, during which time the jammer is least effective. Even when operating against pulse tracking radars to interfere with true target range tracking by the radar, it may be sufficient to produce noise at times at and near the true target's range delay time. The brute force jammer necessarily transmits during all of any one radar transmitter pulse interval time, whereas jamming periods equal to about ten pulsewidths would be adequate. In this case, the power dilution would be equal to ten times the transmitter duty cycle of the radar.

A fourth and even more serious disadvantage of brute force jamming is the fact that no single modulation of the transmitted signal is effective against all modes of angle measurement by the radars. As indicated in the previous chapters, specialized modulation is required for the various search radars as well as for the different forms of angle tracking by tracking radars. Therefore, although the brute force jammer can be very effective in denying the radars the measurement of range to their targets, ability of the radars to measure the target's angular position will not be affected, and, in fact, may be enhanced by the radiation from the jammer.

8.2 RESOURCE MANAGEMENT

The need to use individual and independent modulations against the various radars in the threat environment has dictated a requirement that the ECM system be capable of identifying and separating the signals as they pass through the system modulators. This then led to the capability for effective management of the available power in the ECM transmitter which was not possible with the brute force noise jammer described in the previous section.

Proper management of ECM system resources provides the capability of maintaining effectiveness against two or more victim radars with little compromise in system performance. Sharing of transmitter power is dictated by the requirement for the ECM system to operate in a dense radar environment, and can be accomplished with one or more of the following parameters: frequency, time and space. With any of the power sharing parameters, a dilution of energy available for any one of the victim radars must be accepted as a compromise to system performance.

However, with proper resource management, the effect of the power dilution can be kept to a minimum.

An important requirement for the resource management function is the application of the most effective jamming modulation for each of the victim radars. As indicated in the previous chapters, different types of modulations are required to confuse or deceive the various threat radars expected in the environment. Therefore, managing the transmitter power to ensure an adequate *J/S* for each of the radars is not sufficient; unless the appropriate modulation is included on the transmitted signal, we cannot expect maximum interference on the operation of the radar defense system.

To provide proper management of ECM system resources, the receiver associated with the ECM system must be capable of analyzing all detectable signals impinging on the receiver antenna. Because of the extremely dense radar environment, the receiver must be able to sort out the signals to identify the extent of the threat each radar presents to the survivability of the vehicle or its mission. The resource management system also must be able to control the signals passing through the ECM amplifier system so that only those that need to be transmitted share in the transmitter power available, and that each of signals are appropriately modulated and directed in space.

8.3 RESOURCE MANAGEMENT OF RESPONSES

In this discussion, we assume that the ECM system is operated in a *constant power mode,* which dictates that the maximum power available in the transmitter be used to respond to any one (or all) of the victim radars. If the ECM system is operated in a *constant gain mode,* no such dilution of power results, unless one (or more) of the signals passing through the system forces the transmitter to operate at maximum power.

Sharing of the available resources in the ECM transmitter can be achieved by controlling the signals passing through the transmitter on the basis of one or more of the following parameters:

● frequency;
● time;
● space.

8.3.1 Resource Sharing in Frequency

The carrier frequencies of two or more radars in the threat environment are likely to be more than a few megahertz from each other. An ECM system, in this case, can be designed to separate these radars in frequency to modulate each with a different jamming technique if necessary. If the signals are transmitted simultaneously via the output amplifier, a sharing of the available power results for each victim radar. The amount of power dilution is inversely proportional to the number of simultaneous signals passing through the ECM transmitter.

To provide management of ECM resources against threat radars on the basis of carrier frequency, the system must detect, separate, and identify all signals impinging on the antenna of the receiver associated with the ECM system; as we will indicate later, this is a formidable requirement because of the large number of signals expected in most engagements. The ECM receiver must be able to measure the frequency of the intercepted signals very accurately (less than a few megahertz) if it is intended to set an ECM signal generator to the radar frequency. The signal generator frequency must be equally accurate if successful interference at the radar is expected.

Against the more advanced radars that use carrier frequency agility, the frequency measurement and signal generator frequency set-on must be accomplished with minimum delay (less than a pulse width). Unless this requirement can be met, the ECM system may be forced into a broadband jamming mode with its attendant dilution of ERP.

The constant power transmitter usually generates a power level, which, unless sufficient antenna isolation is provided, will feed a signal back into the ECM receiver that is of equal or greater power than the radar signals being intercepted by the receiver. The ECM transmission therefore must be interrupted so that the receiver can "look through" the jamming to update its data on the intercepted signals. Because the interruption in the jamming yields the radar "free looks" at the target, the unjammed time must be kept to a minimum. Interruptions less than a millisecond at about a 10% duty cycle have been used effectively by modern noise jammers.

Pulse jammers enjoy adequate receiver "look-through" periods because the transmitter is off during the periods between pulse transmissions. Much effort is being expended to develop techniques of "continuous look-through," which will allow updating of the threat data without interruption of the jamming. Look-through on a frequency-to-frequency basis is one form of such improvement.

Because the frequency of ECM transmission is under the control of the receiver and analysis circuits of the ECM system, individual and unique modulations in amplitude, time, or frequency can be applied to each transmission. Modulations especially tailored for each victim radar can be applied to each signal, which will result in maximum interference with the angle and other target parameter measurements of each radar.

8.3.2 Resource Sharing in Time

If the signals from the radars are pulsed or are from a scanning antenna (so that the time of arrival of their signals does not require simultaneous ECM transmission), no ECM transmitter power dilution results. The ECM transmissions for each of the victim radars are interleaved in time so that at no time is more than one signal passed through the output amplifier of the ECM system. The latter requirement is reasonable for low duty cycle radars.

Because ECM operation against CW or high duty cycle radars is almost always performed with the repeater mode at less than maximum transmitter power, no power dilution results from simultaneous operation against multiple coherent radars. Otherwise, the jamming modulation imposed should be such that time interleaving can be achieved; square wave amplitude modulation for angle jamming falls into this category.

We should point out that with time interleaving of the ECM transmitted signals, although no dilution of peak power results, there is power dilution in the transmitter average power, which may be important, especially in the case of CW and high duty cycle radars.

ECM transmitter time sharing for noise jamming can be achieved as shown in Figure 8.1. As shown in Figure 8.1(a), it is intended to generate noise-like signals at two different frequencies. As shown in Figure 8.1(b), this is accomplished by switching the frequency of the ECM signal generator from one frequency to the other at rates which result in noise-like interference at both frequencies. As indicated in Figure 8.1(b), it is only necessary that the frequency of the ECM transmission be within the detection bandwidth of each receiver.

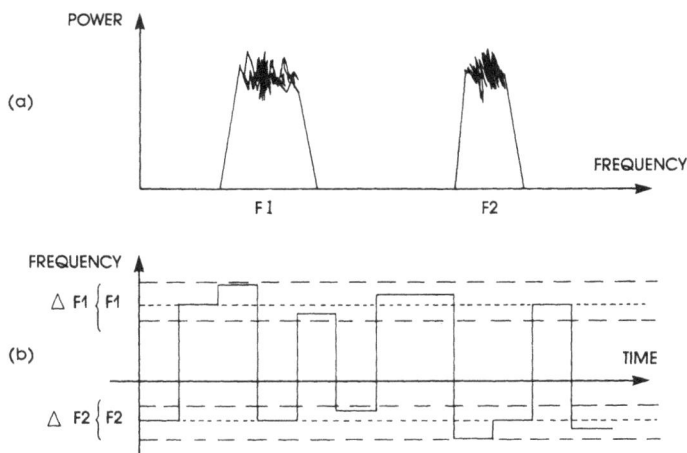

Figure 8.1 Frequency multiplexing.

The switching is performed in a digital fashion; the time duration during each dwell within the receiver bandwidth is commensurate with the pulsewidth, and thus the bandwidth, of the victim radar. We have shown that the effectiveness of the noise jamming does not suffer significantly until the time away from the radar receiver bandwidth is more than three times the time duration within the bandwidth (a duty cycle of 25%). This suggests that simultaneous noise jamming against as many as four victim radars can be achieved with little loss in effectiveness.

Although the switching between frequencies is performed in a noise-like fashion, a pseudo random noise generation technique is used, whereby a unique modulation in amplitude or frequency can be synchronized with each of the frequency transmissions for the purpose of angle or other parameter jamming of each individual radar.

For ECM techniques which require responses on a pulse-to-pulse basis, such as RGPO and multiple targets in range, the ECM system must be able to identify the signals from the victim radar as they enter the ECM receiver so that the appropriate modulation can be applied immediately as it passes through the system. This requirement dictates the ability to anticipate the arrival of the radar pulse on a pulse-to-pulse basis. Receivers have been designed to determine the PRF of these radars so that a gate or window can be generated, where the gate is synchronized with the expected TOA of the radar pulse at the ECM receiver.

8.3.3 Resource Sharing in Space

To operate simultaneously against two or more radars, the ECM can transmit via an antenna with a beamwidth that encompasses the spatial positions occupied by all of the victim radars, or via an antenna which can be steered individually in time to each of the radars. The first approach results in a power dilution equal to the beamwidth of the potential narrow-beam antenna used in the second approach, divided by the beamwidth of the broadband antenna required in the first approach.

Although use of the directional antenna produces a high antenna gain, yielding a higher ERP from the ECM system, average power dilution still results because of the time spent by the steerable antenna in the other radar directions. If the steerable antenna design is such that multiple beams rather than a single steered beam are simultaneously applied against the radars, power dilution results in proportion to the number of beams applied. Nevertheless, for maximum effectiveness, the jamming modulation, in either case, is applied in synchronism with the steering direction.

To provide control of ECM transmissions in space, the ECM receiver design must include the capability to measure the direction of arrival of the signals of interest, and the system must include a transmitting antenna which can be steered in accordance with the measured angles. This dictates the ability to sort and separate these signals for appropriate modulation as they pass through the ECM system.

Because the antenna beam switching is under the control of the ECM receiver and analysis circuits, individual and unique modulations in time, frequency or amplitude can be applied to each of the transmissions as required for individual jamming of the target measurement functions of each victim radar.

Because it is advantageous to synchronize the direction of transmission with the digital noise jamming discussed in the previous section, it is imperative that the speed of beam switching be commensurate with the speed of frequency switching, which is on the order of a microsecond. Electronically steerable antennas are being developed with the required switching speed.

8.3.4 Resource Sharing in Polarization

Because radars operate with different polarizations, it may be possible to isolate ECM transmissions on the basis of this parameter. However, because of the difficulty of measuring this parameter (especially against a radar which uses a receiving antenna different from its transmitting antenna), this parameter has not been applied except in cases where it could be proved that an adequate improvement in J/S could be achieved. An ECM system employing a single fixed polarization suffers a power dilution due to this parameter against all radars except those that receive at that polarization. The dilution which results has not been shown to be significant enough to justify inclusion of polarization steering in ECM system transmitter design. If the victim radar receiving antenna polarization is exactly orthogonal (within 1°) to the ECM transmitting antenna polarization, the power dilution is greater than 35 dB. The polarization can be off by 45° and still be within tolerable limits.

Cross-polarization ECM techniques do require adjustment of the ECM transmitter polarization. These, however, are designed to achieve interference with the target angle measurement by monopulse radars, and require transmission of ECM signals at a polarization which is not favorable to maintaining high effective radiated power.

8.4 RECEIVER REQUIREMENTS FOR RESOURCE MANAGEMENT

As indicated in the previous sections, in order to apply unique and individual modulations on the various signals passing through an ECM system for the purpose of specializing jamming techniques, it is necessary to separate, identify and sort the signals so the appropriate modulation can be applied to each signal of interest. In a typical threat environment, a formidable number of emitters are present, including those used by the enemy and those employed by friendly forces. Although most are no threat or of no interest to the ECM system, all those within the frequency range of the receiver must be identified by the ECM system so that proper action can be taken to reject those of no interest and to determine what priority and what response are most appropriate against those which present a threat to the survivability of the protected vehicle.

The number of signals which are detectable in the receiver is dependent on the sensitivity of the receiver. A receiver that is adjusted to low sensitivity will reduce the number of signals detected in the receiver, but may be too insensitive to detect signals radiated by low power systems which are a direct threat to the vehicle being protected. However, a highly sensitive receiver may force the ECM receiver to detect and analyze more signals than it can adequately process. The requirement to detect signals radiated via the antenna sidelobes of important threat radars demands a highly sensitive receiver (on the order of -100 dBm), which results in an intercepted pulse density on the order of millions of pulses per second in addition to detection of signals from hundreds of high duty cycle and CW radars.

Because the ECM system and its associated receiver must operate over very wide frequency ranges, a severe requirement is placed on its ability to operate individually with unique modulations against multiple threat radars in the environment. The receiver must be able to separate and sort the intercepted signals in such a way that the priority of the threat presented by each can be determined. The receiver must also determine the most appropriate response to be transmitted by the ECM system, as well as the jamming modulations required for each threat radar. The ECM receiver must also control the signals as they enter the receiving antenna to the ECM system so that only those which are required in the response pass through the ECM modulating and amplifying system.

The requirements for an ECM receiver and its processing are aggravated because of the following factors:

- *Very Dense Signal Environment.* It has been shown that the expected environment contains many radars which are associated with the defense systems employed by the adversary.
- *Wide Range of Operating Frequencies.* Because modern ECM systems are designed to operate against a wide variety of radars, their frequency coverage must encompass as wide a frequency range as the component technology will allow. ECM systems are currently being produced to operate over several octaves (2 to 16 GHz, for example).
- *Variety of Emitters.* As indicated in the previous chapters, although the fire control radars are of immediate concern to the ECM system, the probability of survival of the vehicles protected may be increased if jamming against the other radars in the defense system is employed simultaneously.
- *Deliberate Variation of Emitter Parameters.* To aggravate the ECM problem, radar designers have introduced deliberate variation of parameters such as carrier frequency and PRF agility into radar systems.
- *Existence of Friendly and Nonmilitary Emitters.* The ECM receiving antenna is equally receptive to friendly radars located in the environment as well as radars associated with aircraft control at airports located in the area.

8.4.1 Signal Processing

Figure 8.2 is a block diagram of an ECM system employing a digital processor to analyze the signals passed through the various receivers used to detect the different signals expected in the threat environment. Such receivers may include a filter bank receiver or an *instantaneous frequency measurement* (IFM) receiver to measure the carrier frequency of the intercepted signals. Receivers also measure the pulsewidth from the pulse radars and are used to identify the CW radars. An AOA measuring receiver is also required to determine the angular location of the individual emitters.

As shown in the figure, two paths are indicated for the signals entering the receive antenna array. The high-speed, low-latency path is necessary to pass signals with minimum delay through the ECM modulation and amplification sections of the system. The minimum delay is required for RGPO techniques as well as responses using the repeater mode. The high-speed, moderate-latency path is used to perform the processing required to separate, sort and identify the signals so that appropriate control and modulation of the signals passing through the low latency path can be achieved.

Ideally, use of a low-latency path for the latter function also would be desirable, except that sorting and identification of the signals demands a limited length of time to be effective. For example, to determine the PRF of pulse radars at least one *pulse repetition interval* (PRI) is required; in the case of radars using PRF agility, many such PRIs are required. PRF measurement is significant because it is used to determine the TOA as well as the next pulse TOA in generating the gates required in the low-latency path. PRF measurement is also a significant requirement for the pulse deinterleaving process used in signal sorting.

As indicated in Figure 8.3, the signal processor accepts the measured parameters from the various receivers, sorts, and identifies each emitter by comparing the measured parameters to sets of parameters stored in a threat library. A threat situation is determined and compared with a response library which contains the jamming doctrine. The *jamming doctrine* is dependent on the threat analysis and defines the most appropriate jamming response for the threat situation identified. As indicated in the figure, the threat and response libraries are maintained and updated via the inputs at (b), (c), (d), and (e). The response decision determines how to control the low-latency path in the ECM system.

8.4.2 Signal Processor Data Flow

Figure 8.4 shows the data flow in a typical digital signal processor. The "*data word binning*" function is designed to receive the data provided by the various receivers; modern receivers provide this data in digital form, which is compatible with the digital high speed processors being developed. The development of *very high speed integrated circuits* (VHSIC) has accelerated the development of these processors.

The types of data, their bit size and resolution are shown as "pulse descriptor words" in Table 8 1. The "flags" are used to identify special signals, such as those radiated by CW radars. The parameters listed are those which can be measured, almost instantaneously, at the TOA of the signal and within a pulsewidth, if the detected signal is a pulse. The data is sorted into individual bins for any one or all of the pulse descriptor words.

Table 8.1
Pulse Descriptor Words

Parameter	Bits	Resolution
TOA	25	50 ns
Frequency	18	1 MHz
Polarization	16	1°
Amplitude	7	1 dB
AOA	9	1°
Pulsewidth	13	50 ns
Flags	8	—
TOTAL	96	

The more narrow the resolution in a bin, the less likely that multiple signals will exist in the bin. Unfortunately, the more narrow the resolution, the more bins are required, increasing the cost and complexity of the binning process. Furthermore, the more narrow the bin resolution, the more vulnerable is the sorting process for radars which vary their emitter parameters, such as carrier frequency agility. Generally, the bin width is selected as a compromise to the stated factors, which results in data from two or more emitters placed into one or more bins.

The function of the *deinterleaving* process in the figure is to separate signals from multiple emitters in any one of the bins. The deinterleaving process is generally performed with PRF measurement of each of the pulse trains contained within the bin; separation is performed on basis of the TOA and PRF of each of the pulse trains in the bin.

Because the angle of arrival of the intercepted signal is the measurable parameter which cannot be easily varied by an adversary, it is the one which is used as the primary separation parameter. Time correlation of the three parameters—AOA, PRF, and carrier frequency, is used to sort and identify the emitters before a comparison is made in the threat library.

8.4.2.1 Typical Jamming Sequence

A typical sequence of operations for detection, analysis and transmission of ECM signals is as follows:

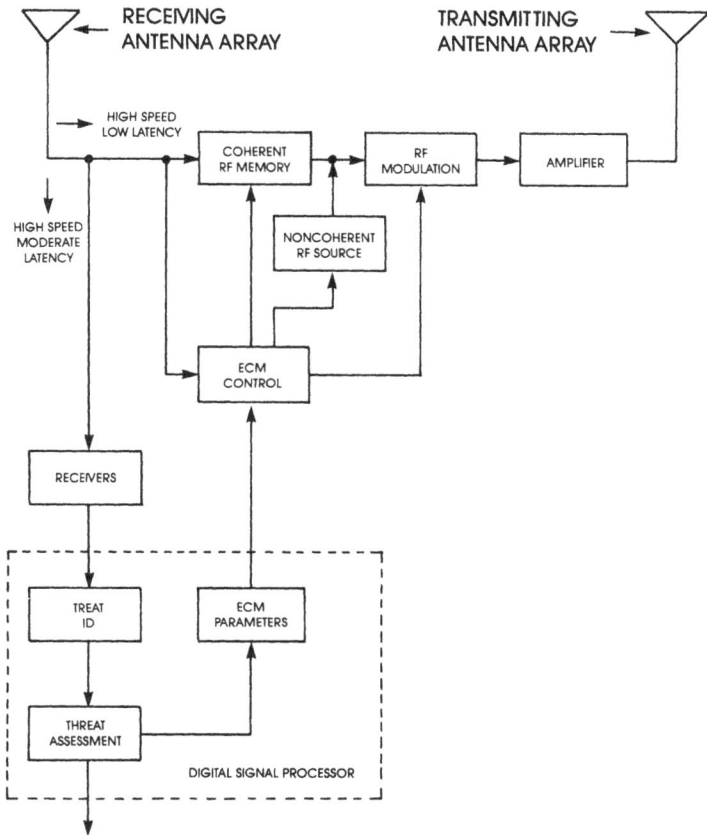

Figure 8.2 Multimode ECM system block diagram.

1. *Measurement of Emitter Parameters.* The various receivers in the system determine measurable parameters of the intercepted signals and pass them to the binning process in digital form, as shown in Table 8.1.
2. *Binning.* The binning process sorts the signals on the basis of one or more of the parameters accepted from the receivers, and places the data in the appropriate bins. The resolution of each bin is generally large enough to ensure that a parameter from one radar is not sorted into two or more separate bins.
3. *Deinterleaving.* For bins containing parameters from two or more emitters, the signals are further sorted by the deinterleaving process on the basis of TOA and pulse interval time. This process becomes very challenging in the case of emitters using PRF agility.

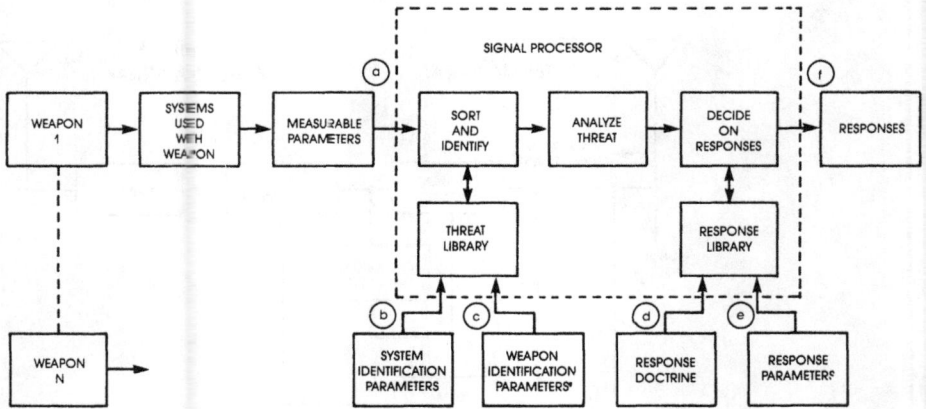

Figure 8.3 Signal processing in multimode system.

4. *Emitter Sorting.* Coordination of all parameters of any one emitter with TOA and PRF sorts the emitters for further processing.
5. *Emitter Identification.* The identified emitters are compared with the threat library to determine the extent and priority of the threat to survival of the protected vehicle.
6. *Response Decision.* Comparison of the threat situation with the response library determines the optimum response in that environment.
7. *Response Parameters Generated.* The modulation parameters required for optimum jamming in the specified threat environment are identified and passed along to the ECM control circuits. In addition, a pulse tracker is generated in synchronism with the measured PRF of each victim pulse radar contained in the response decision.
8. *Jamming Initiated.* The pulse tracker provides gates for the low latency path that identifies the TOA of pulse signals which are to be modulated in the response. The jamming modulation required on each such pulse, on a pulse-to-pulse basis, is applied as it passes through the ECM system. If noise jamming is also included in the response decision, it is interleaved with the pulse transmission, containing whatever jamming modulation is appropriate for that signal. The ECM control ensures that the ECM signals are truly interleaved to prevent simultaneous insertion, and thus power dilution, into the output amplifier.

8.5 SUMMARY

Dedicated ECM systems (those which are designed to operate exclusively against particular types of radars with limited jamming modulation capability) are no longer

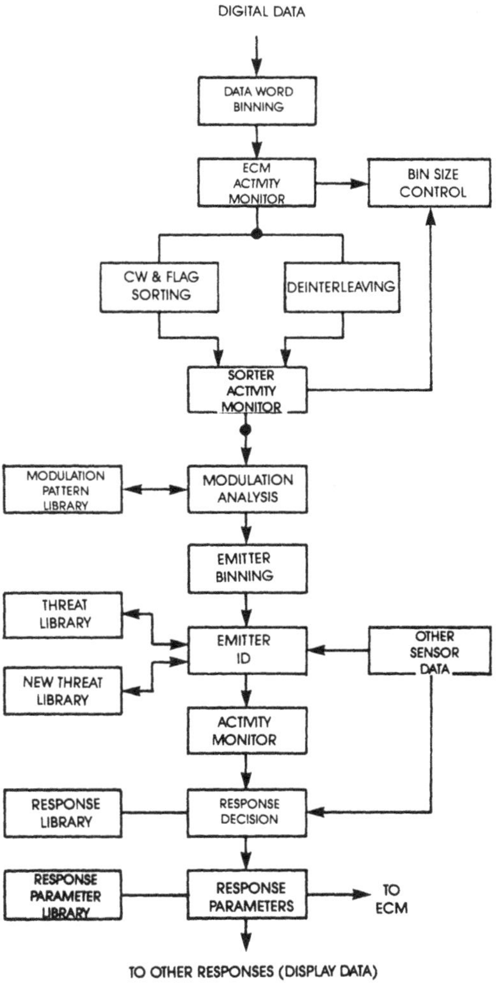

Figure 8.4 Data flow in signal processor.

cost-effective for the protection of vehicles penetrating enemy defenses. The advent of technologies such as VHSICs and *Monolithic Microwave Integrated Circuits* (MMICs) has provided the impetus to the design of multimode multifrequency ECM systems within the cost, weight and space limitations associated with the vehicles.

Excellent receivers have become available which measure the parameters of emitters contained in the threat environment with high speed and accuracy; these parameter measurements are required to identify the radars which are a threat to the survivability of the protected aircraft. High speed signal analysis and decision-making circuits have been developed which require minimum delay for ECM responses. Microwave components have been developed which allow high speed control and modulation of signals passing through the ECM system.

Although brute force jammers can be developed with sufficient power to overcome the power dilution problem, the effect on the radars is primarily in their range measurement capability. Such a jammer can be used in a stand-off vehicle at an angle different from that of vehicles it is designed to protect to provide angular protection for those vehicles. The jammer in the stand-off role must then also provide adequate power to overcome the loss of antenna gain due to operation into the sidelobes of the victim radar receiving antenna.

To provide jamming against the angle measurement function of the victim radars, modulation (in frequency, time, or amplitude), of the ECM transmitted signal is required. Against a single victim radar, this is not a difficult requirement However, when operating in the type of threat environment expected for the penetration vehicles, resource management (as described in this chapter) is required.

Chapter 9
RADAR-ECM CHALLENGES

9.1 INTRODUCTION

The development of radar provided defense systems with the ability to detect and locate vehicles of interest under many adverse conditions, such as night, weather, clouds, *et cetera*. This became especially important in wartime because it prevented vehicles penetrating hostile territories from doing so unobserved even under adverse weather conditions or smoke screens. Furthermore, because the radar signals travel at the speed of light, the radar is able to determine the location of vehicles of interest in small fractions of a second. Because of this extraordinary capability, from the time a radar was first used in warfare, effort was expended by the adversary to find ways of interfering with its operation.

Designers of penetrating vehicles developed methods which would degrade or deny the radars their capability; this operation is known as radar countermeasures, and includes active electronic devices as well as passive devices such as chaff and corner reflectors. Since the first radar-ECM engagement in World War II, the confrontation between radar designers and countermeasures designers has flourished, indeed escalated, and has become an extremely important motivator for new radar design as well as countermeasures system design. In this country, radar designers operate on the assumption that an enemy defense system includes an ECM capability as good, if not better, than ours. Then, too, our ECM designers operate on the assumption that the adversary possesses a radar capability as good, if not better, than ours.

Initially, designers of radars were primarily concerned with interference in the detection process from inherent noise in the receiver circuits, and any external unintentionally generated noise which interfered with the detection of the reflected signals. With the advent of radar countermeasures, it became necessary to design radars to compensate for the intentionally generated interference. Initially, this interference was created with passive devices, such as chaff and corner reflectors; later, electronic jamming devices operating at frequencies which included the carrier frequency of the victim radar were developed. The transmitted signals were modulated to interfere with detection and location of the reflected signal by the radar. The latter development represented the beginning of the systems which are the subject of this book, active electronic countermeasures.

Passive devices are still used very effectively to interfere with the operation of radar systems. There is little reason to believe that their effectiveness will ever be negated with advanced radar designs, especially because improvements in the design and use of the passive devices are continually being made to counter any such advancement in the radar. Nevertheless, passive devices suffer from limitations that can only be overcome with actively generated interfering signals. The interested reader is referred to the literature for more details on passive devices.

Before discussing advances in radar development that are expected to present new challenges to ECM design, we will list developments previously introduced into radar systems to counter ECM devices, and developments in ECM system design which were developed to counter such radar advances. This listing should represent the current scenario from which new developments are expected to emerge.

9.1.1 Radar Developments

The confrontation between radar designers and ECM designers has led to many new developments of radar design techniques and new technologies required to mechanize the techniques. Techniques developed by radar designers primarily to improve the operation of the radar but which affect ECM devices include:

- *Monopulse Angle Tracking.* Although the monopulse technique of target angle tracking was primarily developed to improve the accuracy and reliability of that measurement, it made many of the ECM techniques which were primarily designed against sequential lobing radars ineffective. Monopulse angle tracking required new angle jamming techniques in ECM systems; as indicated in Chapter 6, a reliable ECM technique for an on-board installation has yet to be developed.
- *Coherent Radar Operation.* Coherent techniques have been included in advanced radars primarily to provide a capability of extracting moving targets from a background of clutter generated by ground reflections. This mode of operation made the microwave signal storage systems which were developed to operate against noncoherent radars obsolete.
- *Higher ERP.* Higher power transmitters and higher antenna gains were included in advanced radars to produce a longer range of target detection. As indicated in Chapter 2, higher radar ERP levels have a direct effect on required ECM transmitter power levels.
- *Pulse Compression.* Pulse compression in radars was developed primarily to

achieve higher average transmitter power levels without increasing peak power requirements, and yet preserving desired short pulsewidths required for range resolution of small targets. This development affected the emitter detection capability of ECM receivers and also required coherent microwave storage devices in ECM systems, if false targets were to be generated.

Techniques developed by radar designers primarily to counter ECM systems include:

- *PRF Agility.* This technique was developed to deny ECM systems the capability of generating targets which appear at the radar at ranges of less than that to the vehicle the ECM is protecting.
- *Carrier Frequency Agility.* This technique was developed to counter the frequency set-on capabilities of ECM systems. This antijamming technique prevents the ECM system from generating noise or false targets at radar ranges of less than the vehicles being protected by the ECM system.
- *Carrier Frequency Diversity.* Operating the radar at a wide and diverse set of carrier frequencies forces the ECM system to provide a diversity at least equal to that used in the radar system.
- *Leading-Edge Range Tracking.* Because of the inherent delay in ECM systems, some radars have included the ability to track only the leading edge of signals entering the radar antenna. It is intended to be a counter to RGPO programs. As indicated in Chapter 5, this capability is not included in the radar system without a compromise in performance.
- *LORO Angle Tracking.* Because of the effectiveness of the inverse gain angle jamming technique, which measured the nutation frequency of the radar to produce the required jamming modulations, radar designers developed antennas which nutated only the receiving antenna. This forced ECM designers to develop techniques which did not require radar nutation frequency data.
- *Sidelobe Blanking and Cancelling.* To counter ECM systems transmitting into the radar antenna sidelobes, radar designers have incorporated an auxiliary antenna, which in combination with the main antenna, provided the radar a capability of determining which of the intercepted signals entered the radar antenna via its sidelobes. By identifying the range or doppler frequency of the sidelobe signal, the radar is able to gate out the interfering signals.
- *LPI Transmissions.* To deny detection of the radar signal, spread spectrum techniques are used to yield a radar signal which is below the detection threshold of the ECM receiver. The pulse compression radar technique produces such spectra.
- *Deceptive Transmissions.* By inserting false modulations on its transmitted signal, a radar can deceive or confuse the ECM system as to the correct jamming modulation to impose on its transmitted signals.

These techniques led to the development of new technologies which include:

- *Phased Array Antennas.* These antennas were required to provide flexibility in the beam steering programs in advanced radars. Use of these antennas denied ECM systems the ability to identify the radar antenna scanning program required for effective modulation of the ECM transmitted signal.
- *Ultralow Sidelobe Antennas.* Although typical radar antenna sidelobes were at levels deep enough (> 20 dB) to present a serious ERP requirement on ECM systems, new ultralow sidelobe levels (> 60 dB)make the requirement practically impossible to meet.
- *LORO Antennas.* The LORO antennas were developed primarily to counter ECM systems which were able to measure the nutation frequency of the radar and apply effective interfering modulations on the ECM transmitter signal
- *Frequency Agile RF Generators.* Because frequency agility has a significant effect on ECM systems which attempt to position noise or false targets at near-in radar ranges, the application of these generators was an important development in advanced radars. We should point out, however, that carrier frequency (at least on a pulse-to-pulse basis) is incompatible with rejection of clutter, and hence cannot be used if chaff rejection is also required.
- *Coherent Oscillators.* The capability of extracting the low frequencies associated with doppler processing dictated the development of extremely stable (coherent) oscillators. This development allowed the radar to operate with very low detection bandwidths (< 1 kHz), which, in turn, required coherent operation of ECM systems if they were to compete with true target returns in the radar detection circuits.
- *Master Oscillator Power Amplifier* (MOPA) *Transmitters.* Because magnetrons and similar high-power RF generators cannot easily be operated coherently, low-power coherent oscillators were developed. The MOPA transmitters were developed to provide the amplification required to raise the low power levels of the oscillators to the high power levels required for long-range detection of targets.
- *Coherent TWTs.* To provide high power amplification of coherent signals, amplifiers had to be developed which were free from extraneous modulations in the frequency domain in particular.
- *Synthetic Displays.* The signal processing required in coherent radars is not able to preserve the raw video characteristics of the detected signal, a feature which is used for identification by operators with noncoherent radars.

Although some of these technologies were developed to improve the operation of the radar independent of their effect on ECM, their use has also affected the effectiveness of the ECM systems. For example, the ultralow sidelobe antenna was developed primarily to minimize interference from sidelobe clutter in doppler radar systems. However, among all of the techniques listed above, the ultralow sidelobe antenna has the most significant effect on active ECM effectiveness. The ultralow sidelobe antenna has added truth to the radar designers' contention that if the received signal is of sufficient level to exceed the sensitivity threshold of the radar receiver, the source of the signal (reflected or transmitted), must have been located within the main beam of the antenna.

9.1.2 ECM Developments

Currently, many new techniques and technologies have also been developed by ECM designers to counter the advances made by radar system designers which include:

- *Noise Jammer High Speed Frequency Set-on.* To counter the carrier frequency agility being incorporated into advanced radars, a high speed frequency set-on technique had to be developed.
- *Swept Square Wave (with or without Jog Detection).* The development of LORO and COSRO radars demanded a response from ECM designers. The swept square wave technique was a reasonable compromise accentuated by the technique of jog detection when applicable.
- *Cross-Polarization Angle Jamming.* The introduction of the radar monopulse angle tracking technique presented a formidable challenge to ECM designers, one which still has not been met satisfactorily. The cross-polarization technique is a powerful device, but it suffers from very serious limitations. One of the limitations is the fact that the cross-polarization is in response to measurement of the radar transmitter antenna polarization, which may or may not be that of the radar receiver antenna.
- *Cross-Eye Angle Jamming.* This technique has been found to be effective only when an adequate separation between the ECM receiving and transmitting antennas can be achieved. Because of the required orientation of the antenna separation, the technique is limited to radars that are appropriately located.
- *Data Rate Reduction.* The data rate reduction technique is very effective against a wide variety of angle tracking radars. However, for the technique to be effective, gate stealing techniques (either against the range or doppler trackers), must be effective to achieve the high J/S required.
- *Range (or Velocity) Gate Pull Off with Hold-Out Targets.* These gate stealing techniques are very effective, especially against radars which are run without operator intervention. Generally, these techniques are not too disruptive to interceptor guidance, but they do serve as a means of achieving required J/S for effective angle jamming.
- *Multiple False Targets in Range and Velocity (Doppler).* Multiple false targets are used primarily against radars in the search mode; unless these can also

be generated outside the mainbeam of the victim radar antenna, their effectiveness is questionable.

To support these techniques, many new technologies were developed, which include:

- *Instantaneous Frequency Measurement.* The requirement for high speed set-on of noise jammers spurred the development of instantaneous frequency measurement devices such as the *instantaneous frequency discriminator* (IFD). The IFD can measure the carrier frequency to within a few megaHertz with a single pulse.
- *Voltage Controlled Oscillators.* The VCOs are used as the drivers in MOPA systems, and, in high speed noise jammer frequency set-on, must be capable of being tuned over very wide bandwidths in very short periods of time (on the order of nanoseconds).
- *Master Oscillator Power Amplifier System.* Because the high speed tunable VCOs operate at relatively low power levels (on the order of milliwatts), they are used to drive high power level amplifiers in a MOPA configuration. The MOPA concept is especially useful where jamming using multiple signals and multiple modulations is required.
- *Broadband RF Components.* Because defense radar systems are operated over a very large range of carrier frequencies (several octaves), component manufacturers were forced to extend the bandwidth capabilities of components to meet ECM requirements. Current ECM systems operate over instantaneous bandwidths slightly more than one octave. Extended development was carried out to broaden the operating bandwidth of the following components:
  - Traveling Wave Tubes
  - Low-Noise RF Amplifiers
  - Phase Shifters and Amplitude Modulators
  - Receiving and Transmitting Antennas
  - Electronically Steerable Broadband Antennas
  - Phased Array Antennas
  - Antenna Cross-Polarizers
  - Microwave Filter Banks
- *Noncoherent Microwave Signal Storage Subsystems.* To effect RGPO, a microwave storage system had to be developed to store the intercepted signal (or a reasonable replica), for at least four microseconds. Noncoherent microwave signal memories have been produced to satisfy the requirement, except that as much as 50% of the ECM transmitted power may be outside the detection bandwidth of the radar receiver.
- *Digital RF Memory.* The introduction of coherent radars into the defense

radar environment dictated the need for microwave storage systems which could reproduce the intercepted signal to within the detection bandwidth of those radars ($<$ 1 kHz). Much effort is being expended to develop a digital storage system which provides the required capability.

9.2 FUTURE RADAR-ECM ADVANCEMENTS

9.2.1 Introduction

That the confrontation between radar designers and ECM designers has continued with much vigor is obvious from the foregoing discussion. There is no reason to believe that this conflict will not continue into the future. This is what makes the study of radar techniques and ECM techniques so interesting and fascinating; it not only requires knowledge of the current technology in both areas, but it also requires an ability to anticipate any technique developed by the adversary to counter some new technique that may be developed.

9.2.2 Radar Advancements

There are many radar advancements being studied which are intended not only to improve the ability of the radar to counter unintentional interference, but also to counter intentionally generated interference. In either case, the advancements are expected to increase the difficulty for ECM systems to deny the radars their required capability. Among the more important radar developments are those which follow:

1. *Higher Transmitter Power Levels.* The prime driver for this requirement is to provide the radar the ability to detect smaller radar cross-sectional area vehicles such as missiles, as well as the new stealth vehicles. The effect on ECM systems is on the ERP required, because that factor is directly proportional to the radar transmitter power level.
2. *Electronically Steerable Antennas.* The advent of electronically steerable phased arrays has provided the radar greater flexibility in real-time antenna pattern generation; this technology allows high speed beam steering of antenna gain patterns among several targets, as well as adjustment of the individual beams for maximum detection of targets or nulling of externally generated interference. The use of this antenna destroys the ability of ECM systems to predict future positions of the radar antenna beam, which is a requirement to ensure effectiveness with several of the ECM techniques.
3. *Sensor Integration (Sensor Fusion).* Most radar defense systems are accompanied by other sensors which can be used to complement or supplement the radar system, especially when the operation of the radar is degraded

due to component failure or external interference. Furthermore, with sensor integration, radar defense systems can better identify the nature and intent of the vehicles detected by the radar. This condition requires that the ECM system be accompanied by devices which attack these other sensors.

4. *Multistatic Radar Operation.* This mode of operation positions the radar receivers at locations which are remote from the transmitter. Although this concept is intended to provide greater ranges of detection into hostile territories by placing the receivers nearer to the engagement line, it results in a serious limitation to ECM operation. In many cases, an ECM system is required to direct its transmitted energy toward the radar receiver with a high gain antenna, especially when it is required to transmit into the radar receiver antenna sidelobes. With monostatic radar operation, the radar receiver is collocated with the transmitter, and by determination of the AOA of the detected transmitter signal, the location of the radar receiver is also determined. With bistatic or multistatic operation, location of the radar transmitter is useless in effective direction of the jamming signal because the ECM is intended for the radar receiver and not the transmitter.

5. *LPI.* To provide effective power management of their transmitter energy, modern ECM systems are required to detect and analyze the intercepted radar signals. LPI radar systems are intended to transmit signals which are not detectable at the ECM system, and thus deny it the capability to recognize the presence of the radar in the environment or to decide on the type of countermeasures to be used.

6. *Use of Artificial Intelligence.* Current radar systems possess numerous algorithms in the analysis of detected signals. Although these algorithms are flexible enough that they can easily be reprogrammed to meet new requirements, until they are, the radar system is vulnerable to radar returns or interference signals which were not anticipated when the algorithms were first installed. More advanced radar systems will contain learning algorithms (*artificial intelligence,* AI), which will automatically adjust to the changing environment. This capability will make fragile many of the ECM techniques designed to exploit the vulnerabilities in the logic used by current radars.

7. *Multifunction Antennas.* Newer radar systems will use a single antenna to perform several sensor functions which are currently performed with a separate antenna for each function. This concept may jeopardize effective ECM operation, especially if the radar antenna is designed to include a passive mode of operation. This mode allows the radar to track the radiation from an ECM system without use of the radar transmission. Operating passively on the ECM radiated signal allows the radar to maintain angle track on a signal even during fades of the radar signal or during periods of interference.

8. *Ultra-agile Carrier Frequencies.* Although some current radars can operate with some degree of frequency agility, there are many, especially coherently

operated radars, which are required to maintain constant carrier frequency operation, at least over a coherent detection interval. Because random pulse-to-pulse frequency agility is the most difficult to counter with ECM systems, much effort is being expended to provide this capability even in coherent radars.

9. *Deceptive Transmissions.* Modern ECM system receivers detect and analyze intercepted radar signals to determine the type of radar, its mode of operation, and its intention. In addition, several ECM techniques require data as to the type of modulation the radar expects on the reflected signal in order to effectively impose interfering modulations on the ECM transmitted signal. Radar designers suggest placing modulations on the radar transmitted signal which serve to confuse the ECM receiver analysis circuits. Such modulation includes amplitude modulation of the transmitted signal, which is erroneously detected by an ECM receiver as a nutation frequency.

10. *Intrapulse Modulation.* Intrapulse modulation includes frequency modulation (chirp) or phase modulation which is intended to provide LPI as well as improved detection of radar target signals. This radar modification complicates the RF memory problem for ECM systems. Currently noncoherent memory devices operate only on the leading edge of the intercepted signal (approximately the first 100 to 200 nanoseconds). Against radars which use intrapulse modulation, the ECM transmitter power penalty is equal to the radar processing gain achieved with this type of operation; typically this loss is on the order of 20 dB. Extensive funds are being expended to develop a coherent memory system which can store and coherently reproduce the radar pulse with its intrapulse modulations.

11. *Fingerprinting of ECM Generated Signals.* When confronted with multiple ECM systems, a radar may be confused in its ability to determine the location of the interfering sources because of resolution ambiguities (either in range or angle) due to multiple strobes. Research is underway by radar designers to determine the unique transmission characteristics of ECM systems which can be used to identify and isolate such systems when simultaneously present. A fingerprint may be the unique spurious signal generation by different ECM transmitters.

12. *Ultralow Sidelobe Antennas.* These antennas are presently being used in more advanced radar systems; as indicated previously, this creates a very serious problem for ECM systems. We expect that in the not too distant future, most, if not all, radar systems engaged by ECM systems will possess this capability. This fix may force the ECM system to concentrate its effectiveness into the main beam of the radar, by using more effective mainbeam techniques or by using cooperative jamming from many ECM sources.

13. *Millimeter-Wave Radars.* This radar "fix" is intended to take advantage of the paucity of broadband components in that region of the frequency spec-

trum. This expectation has given impetus to the development of components in that frequency region for future application in ECM systems.

9.2.3 ECM Advancements

ECM designers are continually monitoring the advances being made in radar designs not only in hostile countries but also in friendly countries to better anticipate and develop new requirements for ECM techniques and advancements in required technology to support these new techniques. The radar advancements listed here are an indication of future requirements for ECM, and have provided the impetus for new developments in ECM design and technology. Several of the more important are listed below:

1. *More Efficient Power Management.* Due to the limited power available for ECM transmitters, especially in airborne applications, it is imperative that efficient use be made of the power radiated from ECM systems. It is important that the power transmitted be within the detection bandwidth of the victim radar, be transmitted into the region that includes the victim radar, and at the time which provides maximum interference with the radar's target signal. Furthermore, it is most important that the appropriate modulations be applied to the transmitted signal to confuse or deceive the radar measurement functions. Although much effort has been expended in this area, much has yet to be done, especially because of the new advancements being incorporated in radar designs.

2. *Stealth.* Although protection of stealth vehicles will be alleviated because of the lower radar cross section, any ECM system which may be aboard these vehicles will be required to operate in a manner that prevents radiation from the systems from betraying its presence or location. It may become necessary that any ECM system used to protect the stealth vehicle be separated from the vehicle before it is energized. Expendables fall into this category.

3. *Sensor Integration.* As with the radar, integration of the data from other sensors collocated with the ECM system can improve the analysis of the threat situation and selection of appropriate responses. Other sensors, which may be integrated with the ECM system, include the radar, infrared, and optics.

4. *Artificial Intelligence.* Currently, the logic being used in ECM systems is restricted to fixed algorithms which can become obsolete with the advanced radars. The ability to learn and to adjust to changing situations must be included in the future design of ECM equipment. Much effort is being expended in designing learning algorithms (artificial intelligence) which may be applied to the ECM problem.

5. *Multifunction Antennas.* To maximize newer antennas, we suggest that multiple use of a common antenna be developed. One study has shown that a phased array radar antenna could be used to provide the functions required by an ECM system, although over a restricted frequency bandwidth. The design of broadband multifunction antennas can help to alleviate this limitation.

6. *Active Element Arrays.* Developments are underway to install low-power amplifiers in each of the elements of a phased array antenna. Inclusion of these active elements in the array provides a high degree of flexibility in the operation of the antenna. With enough of these elements, a higher ERP level can be achieved than with a single-input, high-power transmitter. Furthermore, the phase and amplitude controls on each of these elements yield a high degree of antenna beam control.

7. *Software Programmable.* Although a high degree of computerization has been applied to current ECM systems, the ability to make them completely software programmable to meet new requirements has yet to be realized. ECM systems must be designed so that the hardware produced has little probability of becoming obsolete because of new advances in radar design. Although programming of the analysis logic and modulation parameters is adequate to satisfy current requirements, this effort must be continued to meet the advances made in the radars.

8. *Integrated Distributed ECM.* This is a topic that has received little attention thus far in ECM system applications, but it is one which can lead to extremely effective usage of ECM resources in an area of conflict. Currently there is very little, if any, coordination between the various ECM systems which may be within the same hostile environment, especially in aircraft installations. Generally, each such system operates autonomously, and can result in all systems in an environment operating against the same victim radar and none against a radar which may be a direct threat to the vehicles being protected. One method of integration is to employ a stand-off vehicle, such as the AWACS, to monitor the environment within which the vehicles to be protected are located, and to determine those that ought to be reacting to which radars and with which type of technique.

9.3 SUMMARY

We have tried to make it obvious that the confrontation between radar designers and ECM designers is never-ending. We can only expect that it will become more vigorous in the future. Many capabilities that have not been realizable in previous installations are being planned in both disciplines because of promising new technologies.

The development of VHSIC and MMIC technologies permits inclusion of very flexible and sophisticated analyses of signals, as well as microwave signal processing, which heretofore have been excluded from either radar or ECM systems, where space weight, and power have been limited.

Artificial intelligence will have a significant effect on future radar and ECM designs. Whereas current systems have evolved into a confrontation between the designers of the two types of equipment, we can expect that future equipment may generate a confrontation between artificial intelligence designs.

Appendix A
Use of the Decibel

A.1 INTRODUCTION

Throughout this book (and in the industry), much use is made of the decibel (dB) to describe performance and relationships among the various parameters used in electronic warfare systems. This is because the performance capabilities of many of the components used in EW systems are described in decibels (i.e., the gain of amplifiers, loss in attenuators, gain of antennas, sensitivity of receivers). The decibel has been applied to all ratios of parameters, such as the jamming-to-signal ratio and the radar cross section of vehicles relative to some reference value, even though the nomenclature was intially applied to the ratio of the power levels of sound. Nevertheless, the parameter values of some EW components are still described in absolute terms, such as transmitter power levels in kilowatts. In the latter case, we have found converting these values to decibels to be more beneficial for use in equations with the other parameters.

Why use decibels rather than absolute values? As we will show later, the decibel is a special logarithmic value that leads to numerous advantages by its use. When all values of the parameters in even the most complex equation are given in decibels, the solution often requires merely simple addition and subtraction of these values and, at most, multiplication of them when powers and roots are included. The more significant advantages gained with use of decibels are given below:

1. *Multiplication of Values:* The multiplication of several large absolute values is performed with simple addition of their decibel equivalents. When the equation consists of absolute values and decibel values, converting the absolute values to their decibel equivalents is advantageous.
2. *Division of Values:* The division of several large absolute values is performed with simple subtraction of their decibel values. When the equation consists of absolute values and decibel values, converting the absolute values to their decibel equivalents is advantageous before solving the equation.

3. *Value raised to a Power:* A value raised to any power is performed by multiplying its decibel equivalent by the power value, and then finding the antilog of the result.

4. *Root of a Value:* Any root of a value can be easily determined by dividing its decibel equivalent by the root, and then finding the antilog of the result.

5. *Graphing:* The most significant advantage gained with use of the decibel is for graphing the functions required in electronic warfare formulas. This is especially so when the range of values is several orders of magnitude, such as for deriving the radar range to its targets and the power levels detected in radar receivers as the target range changes.

A.2 GRAPHING

Typical radar-to-target ranges employed in radar-ECM calculations are usually over three decades (e.g., one to a thousand kilometers). On a linear scale, the coordinate would need to show intervals over a scale of one thousand. Using the logarithmic scale, the ratio of 1000:1 results in intervals over a range of only 30 which is a more reasonable requirement for graphing. In addition, as shown in the radar equation, the radar received power level varies by a factor of 10,000 (40 dB) when the target range varies by a factor of 10. Using a logarithmic scale for that parameter results in an interval range of 40 instead of 10,000. This is illustrated in the log-log graph in Figure A.1; the y-coordinate indicates the received power level in logarithmic coordinates and the x-coordinate shows the target range in logarithmic coordinates. The figure also shows the decibel value of the radar received power. With the use of the decibel value rather than the absolute value, linear coordinates can be used for the y-coordinate; this is justified by the fact that received power levels are most often expressed in decibels.

A more significant advantage of using logarithmic graphing procedures is that the solution of the range and ECM equations as well as *J/S* values, *et cetera,* plot as straight lines on the graph. Were we to plot such equations on linear graphs, we would need to determine an unreasonable number of points to establish the graph of all points on the curve. Figure A.2 shows such a plot for the simple $y = x^2$ equation. However, when the equations are plotted on logarithmic graphs, a straight line solution results. In this case, only two points need to be plotted to determine the location of the line. With these two points, we can draw a line that provides the solution of the equation at every point.

Assuming an equation of the form:

$$y = Bx^s$$

and finding the logarithms on each side of the equation:

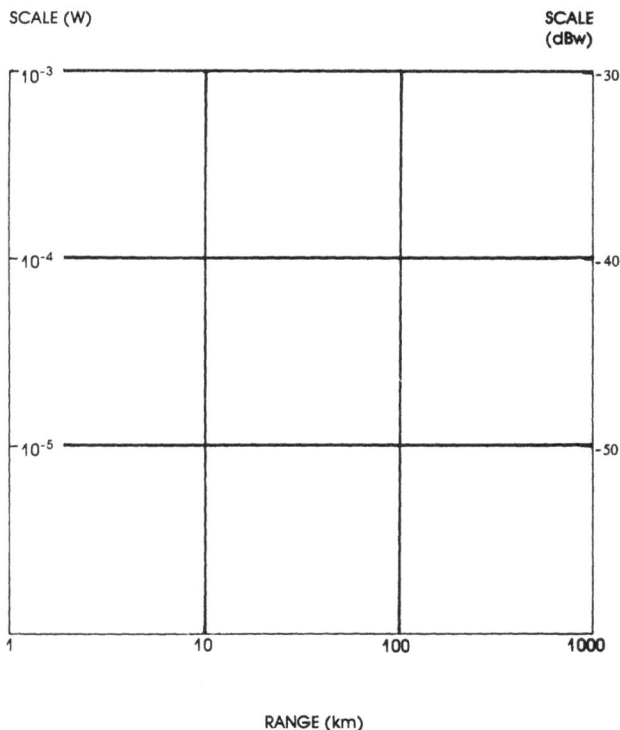

Figure A.1 Log-log graph.

$\log y = \log B + s \log x$

On logarithmic charts, the logs are linear, so that

$Y = K + s \cdot X$

on the log-log graph, because

$Y = \log y$
$K = \log B$
$X = \log x$

Using analytic geometery, we can show that this equation is graphed as a straight line, where K is the Y intercept point, s is the slope of the line, and X is the x-coordinate. The equation:

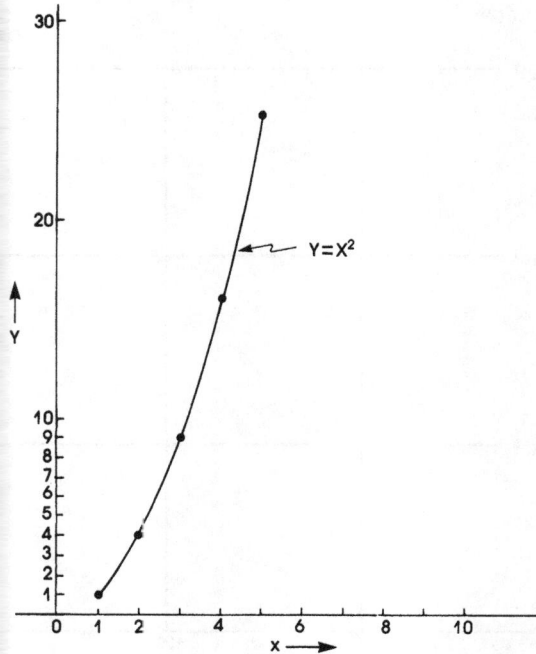

Figure A.2 Linear scale example.

$$y = x^2$$

is shown plotted in Figure A.3 as a straight line.

A.3 DECIBEL DETERMINATION

The decibel was first suggested for the measurement of relative sound levels as follows:

$$\text{Number of bels} = \log(P_2/P_1)$$

where P_2 is one sound power level and P_1 is the other, usually a reference level

However, as was later determined, the ratio of power levels usually experienced in practice resulted in small fractions of bels. This suggested that the bel was too large a measure for the power levels experienced. Researchers then decided to use the decibel as the unit of measure, where

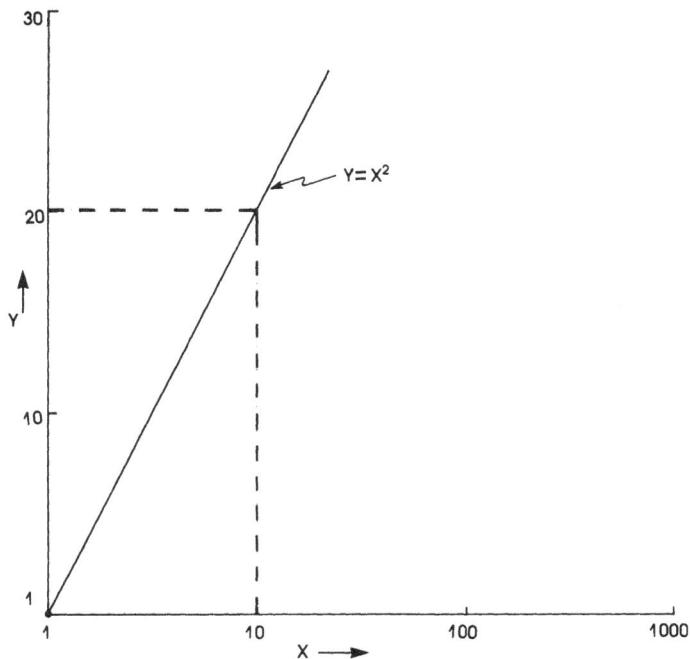

Figure A.3 Log-log scale example.

1 decibel $= 1/10$ bel

or

10 decibels $= 1$ bel

Therefore,

Number of decibels $= 10 \log(P_2/P_1)$

Although the decibel was originally defined for relative sound levels, the more general usage, except for voltage ratios, is

Number of decibels $= 10 \log(\text{any ratio})$

The ratio is often relative to some standard value for P_1, such as one milliwatt or one watt. Some of the more important ratios are

dBm = dB relative to one milliwatt (receiver sensitivity);
dBW = dB relative to one watt (transmitter power levels);
dBi = dB relative to isotropic gains (antenna gain);
dBm^2 = dB relative to one square meter (cross sections);
dB = dB not relative to any necessarily standard value (amplifier gains).

Table A.1 lists the approximately decibel value for all ratios from 1 to 10. Using this table, we can quickly determine the approximate decibel value of all numbers (this is demonstrated in Section A.4). We should observe that all ratios of 2:1 result in a change in decibel value of 3 dB because the log of 2 is approximately 0.3, which, when multiplied by 10, equals 3. Any ratio of 2, such as 2000:1000, 30:15, represents a 3 dB change in value. We can see in the table that the ratio of 4 is 3 dB greater than that at 2, and the ratio of 8 is 3 dB greater than that at 4. A ratio of 4 is 6 dB (4:1), and a ratio of 8 (8:1) is 9 dB, a 3 dB change for each doubling of the ratio.

Table A.1

Ratio	Decibels
1	0
2	3
3	5
4	6
5	7
6	8
7	8
8	9
9	9

A.4 CONVERSION FROM VALUE TO DECIBELS

To convert any ratio to its decibel value, we use the following steps:

1. Move the decimal point to its unit position.
2. If the move is to the left N places, let $A = +10 \cdot N$.
3. If the move is to the right N places, let $A = -10 \cdot N$.
4. Convert the number in the units position to its decibel value, using the table if necessary.
5. Add this value to A, to get the final decibel value.

Example:

$$21312.47 = 2.131247 \cdot 10^4 \qquad \text{(Step 1)}$$
$$N = +4$$
$$A = +10 \cdot 4 = 40 \qquad \text{(Step 2)}$$
$$10\log 2 = 3 \qquad \text{(Step 4)}$$
$$40 + 3 = 43 \qquad \text{(Step 5)}$$

Therefore,

$$\log 21312.47 = 43 \text{ dB}$$

This result compares with a rigorous determination of the logarithm for the number in the example, which is 43.3.

Another example:

.00402894

In this case, $N = -3$ and $A = -30$; the units value is 4, the 10log of which is 6. The log of the number is $6 - 30 = -24$ dB. This compares with the rigorous value of -23.7.

A.5 CONVERTING DECIBELS TO RATIO

To convert a decibel value to its approximate ratio we use the following steps:

1. Round off the value to its nearest unit value if a decimal exists.
2. Convert the units digit to actual ratio, using the table if necessary.
3. Add a number of zeroes to the right-hand side equal to the tens digit.
4. If the decibel value is negative, take the reciprocal of the ratio determined in step 3.

Example:

    Decibel value = 36
    Antilog of units value (6) = 4   (Step 2)
    Tens digit = 3
    Add 3 zeros to the value 4   (Step 3)

Therefore,
    antilog 36 = 4000

    for decibel value = -36
    Antilog $(-36) = 1/4000 = .00025$ (Step 4)

Appendix B
Who Sees Whom First?

B.1 PROBLEM

In electronic countermeasures systems, knowing whether a victim radar can detect the presence of a vehicle carrying a protective ECM system before its associated receiver can detect the radiations from the radar is advantageous, and at times critical. For some confusion ECM techniques used against the search function of a radar (such as false targets and noise) as an imperative, the ECM must become operative before the radar can detect the aircraft, but not too long before this occurs. For other techniques, such as track-breaking countermeasures, the ECM system must remain off the air until we can determine that radar tracking has begun. Otherwise, the ECM may prematurely alert the victim radar and afford it a greater range of initial detection.

This appendix is provided to define the conditions required to ensure that an ECM receiver aboard the vehicle can detect the signals of a victim radar before the latter can detect the signals reflected from the aircraft. In the case of track-breaking countermeasures, although the radar signal may be detected before the radar detects the aircraft, the ECM can be held off until actual tracking of the aircraft by the radar is recognized by the ECM receiver.

B.2 APPROACH

This study was made by using the radar range equation to determine the range of detection of the radar and the one-way Friis transmission formula to determine that of the ECM receiver. To simplify this determination, the two detection ranges are normalized to what is called a crossover detection range, r_c, being the range at which both the radar and the ECM receiver detect simultaneously, based on a fictitious, but calculated, value of radar power, P_c, which is defined as the crossover power level. As shown below, this value is a function of the sensitivity of both the radar and ECM receiver.

A comparison of the actual radar transmitter power with the calculated cross-over power level will determine whether the radar or the ECM receiver has the detection advantage.

B.3 CONCLUSIONS

The study showed that either the radar or ECM receiver can detect first, depending on the radar transmitter's power level and the difference between the detection levels of the radar and ECM receivers.

To determine quickly whether the radar or ECM receiver has the detection advantage, we need only solve for P_c, the crossover power level, as follows:

$$P_c = \frac{4\pi\sigma}{G_j^2 \, \lambda^2} \frac{S_j^2}{S_r} \tag{B.1}$$

where

$$\sigma = \text{RCS of vehicle,}$$
$$G_j = \text{ECM receiver antenna gain,}$$
$$\lambda = \text{radar carrier wavelength,}$$
$$S_j = \text{detection threshold of ECM receiver,}$$
$$S_r = \text{detection threshold of radar.}$$

The equation is based on the assumption that radar antenna gain on reception is equal to the radar antenna gain on transmission. If the gains are not equal, the equation must be multiplied by the ratio of the radar receiving antenna gain to the transmitting antenna gain.

Detection threshold is defined as that level of signal at which the detection circuits of the radar or ECM receiver are able to identify reliably the required parameters of the received signals. Simple detection of the presence of an echo signal may not satisfy this criterion.

If the actual transmitted power of the radar of interest, P_t, is greater than the calculated crossover power level, P_c, the ECM receiver has the detection advantage. If P_t is less than P_c, the radar has the detection advantage.

Furthermore, we can easily determine the range advantage from the following equation:

$$\frac{r_j}{r_r} = \left(\frac{P_t}{P_c}\right)^{1/4} \tag{B.2}$$

where

r_j = range at which ECM receiver detects radar;

r_r = range at which radar detects the vehicle.

To determine the actual detection ranges, we must first define the range where simultaneous detection will take place, r_c, as follows:

$$r_c = \left[\frac{G_t}{G_j} \times \frac{\sigma}{4\pi} \times \frac{S_j}{S_r} \right]^{1/2} \tag{B.3}$$

where the parameters are as listed above, and

G_t = radar transmitter antenna gain

The relationship between the actual detection ranges and the crossover range, r_c, can be found from

$$r_j = r_c \times \left(\frac{P_t}{P_c} \right)^{1/2} \tag{B.4}$$

$$r_r = r_c \times \left(\frac{P_t}{P_c} \right)^{1/4} \tag{B.5}$$

B.4 EXAMPLE

Assume the following parameters for the radar:

P_t = 1 MW
G_t = 30 dBi
λ = 10 cm
S_r = −90 dBm

for the ECM receiver:

G_j = 0 dBi
S_j = −25 dBm

and for the aircraft:

$$\sigma = 20 \text{ m}^2$$

Therefore, using equation (B.1), we have

$$P_c = 84 \text{ dBm} \rightarrow 250 \text{ kW}$$

and

$$\frac{r_j}{r_r} = 1.414$$

The detection range of the ECM receiver is therefore 1.414 times that of the radar receiver. Furthermore, the crossover range is

$$r_c = 48.5 \text{ dBm}$$
$$r_c = 70 \text{ km}$$

Then, by using

$$r_r = r_c \times \left(\frac{P}{P_c}\right)^{1/4} = 96 \text{ km}$$

$$r_j = r_c \times \left(\frac{P}{P_c}\right)^{1/2} = 140 \text{ km}$$

Under these conditions, the detection range of the radar is 96 km, whereas the detection range of the ECM receiver is 140 km.

If we now assume that the radar transmitter power is 100 kW, we find that

$$\left.\begin{array}{l} P_c = 250 \text{ kW} \\ r_c = 70 \text{ km} \end{array}\right\} \quad \text{(as before)}$$

$$\frac{r_j}{r_r} = 0.80$$

$$r_r = 55 \text{ km}$$
$$r_j = 44 \text{ km}$$

In this example, the detection range of the radar is 55 km, and that of the ECM receiver is reduced to 44 km. Thus, tenfold reduction in the radar power results in first detection switching from the ECM receiver to the radar. Note, however, that in the latter case both detections occur at a range of less than the calculated crossover range. This is because, in order to allow the radar first detection, its transmitter power level must be so low that radar detection is limited to a range of less than the crossover range.

In the example, a − 25 dBm receiver was postulated for the ECM system to show the effect of the radar detection sensitivity and ECM receiver sensitivity on which system detects first. Modern ECM receivers can be expected to provide sensitivities as good as − 70 dBm or better.

B.5 SUMMARY

As indicated in (B.2), the radar receiver has the detection advantage only if the radar transmitter power level is less than the calculated crossover power level. Therefore, to preserve this detection advantage, the radar transmitter power level is required to be low, or alternatively the sensitivity of the radar receiver to be made much greater than that of the ECM receiver.

As we can see in (B.1), the better is the sensitivity of the radar receiver, the higher is the crossover power level. Thus, the radar transmitter power level can be higher and still preserve the radar detection advantage.

As shown in the example, the crossover range is that at which first detection switches from the radar to the ECM receiver. At ranges of less than this value, the radar detects the target before the ECM receiver can detect the radiation from the radar. At ranges greater than this value, the ECM receiver will detect the radiation from the radar before the radar can detect the reflection from the vehicle.

Equation (B.3) indicates that the crossover range is dependent on the ratio of the detection threshold of the ECM receiver to the detection threshold of the radar receiver. In decibel form, this ratio is defined as the difference in the two detection thresholds. Figure B.1 is a plot of the equation, where the x-coordinate is the crossover range and the y-coordinate is the difference in the two detection thresholds expressed in decibels for vehicle cross-sectional areas of 20 and 100 m^2. For example, for a 20 m^2 target and a sensitivity difference of 50 dB, the crossover range is approximately 13 km. Therefore, if the radar receiver threshold is − 100 dBm, the ECM receiver threshold need be no better than − 50 dBm to ensure first detection outside the 13 km range.

If the ECM application dictates that it detect first outside a 50 km range, we can determine from Figure B.1 that the ECM receiver threshold need not be more than 63 dB greater than the radar receiver threshold. Even if the radar receiver has a sensitivity of − 110 dBm, the ECM receiver need not be better than − 47 dBm to ensure first detection into that range. This level of sensitivity is readily achieved in modern ECM receivers.

We should note that the crossover range, as defined in (B.3), is independent of the radar transmitter power level. A very low radar transmitter power level, as defined by (B.1), will allow the radar the capability to detect the vehicle before

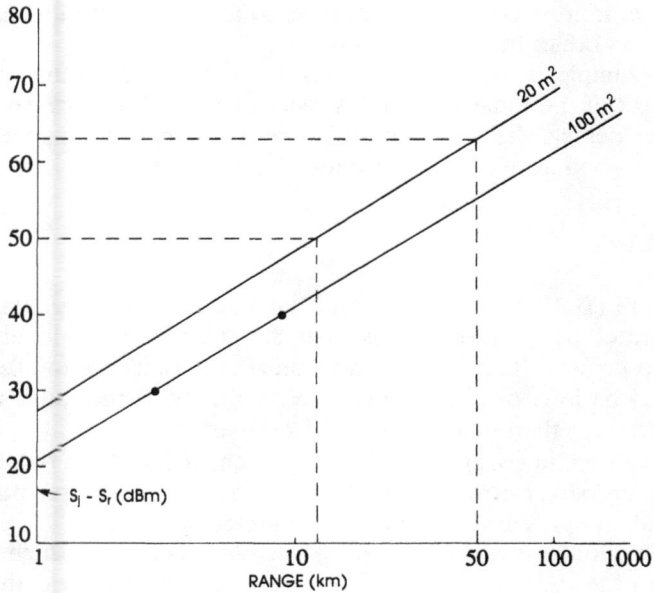

Figure B.1 Crossover range data.

the ECM receiver detects the radar, but this first detection must be at a range of less than the crossover range defined by (B.4). Assuming that the radar transmitter power level is greater than the crossover power level, the ECM receiver will always detect first beyond the corresponding crossover range.

Appendix C
Range Ambiguity Resolution in Pulsed Doppler Radars

C.1 GENERAL

In a conventional pulsed radar, range is measured as a linear function of the time required for a signal to return to the radar after being reflected off the target. The pulse repetition frequency of the radar is made low enough so that the reflected signal arrives at the radar before the next transmitted pulse is initiated. This favorable arrangement is not possible in the case of a pulsed doppler radar, where other factors dictate the use of a much higher PRF, on the order of perhaps 100 kHz, resulting in an unambiguous range on the order of 1 nmi. This ambiguity can be resolved by the use of more than one PRF by the radar. This appendix discusses the means by which true range can be determined with a pulsed doppler radar using multiple PRFs in conjunction with the Chinese remainder theorem.

If more than one target return is present in the detection circuit of the radar, an additional ambiguity is introduced into the measurement of range. This phenomenon is also discussed here, and is referred to as "range ghosts." This phenomenon is a threat to a pulsed doppler radar only if the additional targets are present *within the detection circuit of the radar.* Doppler radars characteristically have two powerful methods for rejecting all targets other than the target of interest: *angle discrimination* and *velocity discrimination.* To be visible to the radar, the unwanted target must be present within the antenna beamwidth of the radar, which means that the unwanted target must be located at the same angle (azimuth and elevation) as the true target. The second means by which a doppler radar discriminates against unwanted targets is the velocity filter. A doppler radar generally has velocity discrimination so precise that it can reject targets with doppler frequencies displaced by greater than 200 Hz (as a typical value) from the frequency of the true target. A doppler frequency of 200 Hz represents, at X-band, a velocity differential of approximately 7 mi/hr. Thus, to be seen by the radar, an unwanted target would need to be at the same angle in space and at the same velocity within 7 mi/hr.

As we have discussed, the criteria for entry into the detecting circuit are difficult for other targets to meet. The additional targets must be present at the same angle, at an almost identical velocity, but at a different range than the true target. The requirement for being at a different range means that the unwanted target must be separated from the true target by a distance at least as great as the range resolution capability of the radar. This spacing, in time, might be typically 0.5 to 1 μs, the radar range equivalent of 250 to 500 ft. A practical situation in which unwanted targets might meet the criteria would be the case of formation flying. Some of the aircraft in the formation might be spaced from others by the requisite amount, and conceivably their velocity differential might be as low as the required 7 mi/hr.

Another situation that may cause unwanted targets to appear within the detection circuit of the radar is where the unwanted targets are generated synthetically. Although ghost targets do not occur frequently in pulsed doppler radars, the fact that they can occur under some circumstances warrants an analysis of the phenomenon.

C.2 TWO-PRF CASE

C.2.1 Range Calculation

The general solution for the problem of determining range by using multiple PRFs involves the use of congruences and the application of the Chinese remainder theorem. By way of illustration, consider the use of two PRFs and one target. The situation is shown in Figure C.1. The PRFs are selected so that the interpulse intervals are relatively prime (i.e., they have no common factors). Because they are relatively prime, the intervals will coincide at some point in time, as shown in Figure C.1.

The question arises as to the units in which τ_1 and τ_2 are to be measured. The interpulse spacing is measured in terms of the size of the resolvable elements which the spacing comprises. For example, if the range resolution of the radar is 1 μs, τ_1 and τ_2 will be measured in microseconds and the computed range will be given in microseconds. If, however, the resolution of the radar is 0.1 μs, intervals will be measured in tenths of microseconds, and the computed range will be given in tenths of microseconds. Each PRF has an associated target return which occurs at a time equivalent to the true range from the transmitted pulse that produced the return. In Figure C.1 this time is T_0 for PRF 1, and is also T_0 for the other PRF. The radar cannot measure T_0, however, because it does not have a record of the time of coincidence of the PRFs. The radar can nonetheless measure the time t_1, which is the interval between the target return and the nearest PRF pulse.

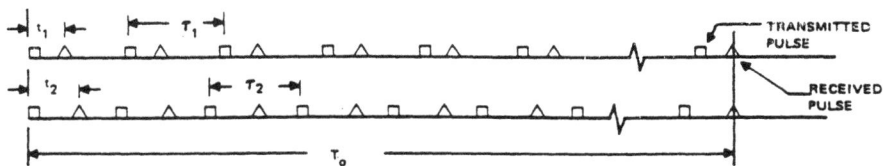

Figure C.1 Time relationships in the two-PRF case.

The radar can also measure t_2, which is the equivalent interval for the other PRF. The problem in computing the time T_0 as equivalent to actual range is to relate T_0 to t_1 and t_2. We know that T_0 is equal to an integral number of $\tau_1 + t_1$. Likewise, T_0 is equal to an integral number of $\tau_2 + t_2$. Expressed mathematically, this is

$$
\begin{aligned}
T_0 &= k\tau_1 + t_1 \\
    &= l\tau_2 + t_2
\end{aligned}
\tag{C.1}
$$

by the theory of congruences [1]. Equation (C.1) can be expressed as

$$
\begin{aligned}
T_0 &\equiv t_1 \ (\mathrm{mod}\ \tau_1) \\
    &\equiv t_2 \ (\mathrm{mod}\ \tau_2)
\end{aligned}
\tag{C.2}
$$

The Chinese remainder theorem states that, if τ_1 and τ_2 are relatively prime by pairs, T_0 can be determined from (C.2) as follows:

$$
T_0 \equiv (t_1)\,(X_1)\,\frac{\tau_0}{\tau_1} + (t_2)\,(X_2)\,\frac{\tau_0}{\tau_2} \ (\mathrm{mod}\ (\tau_1)(\tau_2))
\tag{C.3}
$$

where

$$
\tau_0 = (\tau_1)(\tau_2)
$$

$$
\frac{\tau_0}{\tau_1}\,X_1 \equiv 1 \ (\mathrm{mod}\ \tau_1)
\tag{C.4}
$$

$$
\frac{\tau_0}{\tau_2}\,X_2 \equiv 1 \ (\mathrm{mod}\ \tau_2)
\tag{C.5}
$$

Linear congruences in (C.4) and (C.5) can be solved by using the method described by Ore [2].

C.2.1.1 Interpulse Intervals Are Not Consecutive Integers

As an example of typical parameters, the relatively prime interpulse spacings will be taken as 21 and 19 µs, corresponding to PRFs of 47,619 and 52,631 pulses/s, respectively. Due to the nature of the problem, we obtain

$$T_0 = t_1 \ (\text{mod } \tau_1)$$
$$= t_2 \ (\text{mod } \tau_2)$$

where

$$\tau_1 = 21 \ \mu s$$
$$\tau_2 = 19 \ \mu s$$

Let

$$\tau_0 = (\tau_1)(\tau_2) = (21)(19)$$

Then,

$$\frac{\tau_0}{\tau_1} = 19$$

and

$$\frac{\tau_0}{\tau_2} = 21$$

Let

$$X_1 \frac{\tau_0}{\tau_1} = 1 \ (\text{mod } \tau_1)$$

This is equivalent to

$$19X_1 - 21U = 1 \tag{C.6}$$

where U is an integer. Equation (C.6) is a linear congruence which can be solved by the technique discussed in the three-PRF case in Section C.3.1. By using this technique, we have

$$21 = 19 \cdot 1 + 2$$
$$19 = 2 \cdot 9 + 1$$

1				$10 = 1 \cdot 9 + 1$	
9				$9 = 9 \cdot 1 + 0$	

$$\overline{\quad} \quad \overline{\quad} \quad \overline{\quad} \quad \overline{\quad} \quad \overline{\quad} \quad \overline{\quad}$$

$$1$$
$$0$$

$X_1 = +10, 31, \text{ et cetera}; U = +9$

In a similar manner, let

$$X_2 (21) \equiv 1 \bmod 19$$

or

$$21 X_2 - 19U = 1$$

These coefficients are the same as in the previous case for X_1, so that

$$X_2 = -9, +10, \text{ et cetera}; \quad U = -10$$

By using the Chinese remainder theorem, we can write

$$T_0 \equiv X_1 \frac{\tau_0}{\tau_1} t_1 + X_2 \frac{\tau_0}{\tau_2} t_2 \ (\bmod \ 21{:}19)$$

By using positive values for X_2, we have

$$T_0 \equiv (10)(19) \ t_1 + (10)(21) \ t_2 \ (\bmod \ 399)$$
$$\equiv 190t_1 + 210t_2 \ (\bmod \ 399) \tag{C.7}$$

Equation (C.7) can be rewritten,

$$T_0 \equiv 10 \ \tau_2 t_1 + 10 \ \tau_1 t_2 \ (\bmod \ \tau_1 \tau_2) \tag{C.8}$$

Equation (C.8) is the equivalent of saying that $T_0 = 10 \ \tau_2 t_1 + 10 \ \tau_1 t_2$. The congruence is also solved if the value for T_0 is increased by $\tau_1 \tau_2$ because of the modular nature in which the solution repeats each $\tau_1 \tau_2$ μs. If the quantity in the brackets of (C.8) is negative, the solution has no physical meaning; a physically meaningful solution is obtained by adding multiples of $\tau_1 \tau_2$ until the result is positive.

Equation (C.8) can then be rewritten as

$$T_0 = \frac{10}{f_2} t_1 + \frac{10}{f_1} t_2 + U \left[\frac{1}{f_1} \cdot \frac{1}{f_2} \right]$$

(C.9)

where

$$\frac{1}{f_1} = \tau_1 \text{ and } \frac{1}{f_2} = \tau_2$$

f_1 and f_2 are PRFs 1 and 2, respectively.

Rewriting equation (C.9), we obtain

$$T_0 = \frac{10 f_1 t_1 + 10 f_2 t_2 + U}{f_1 f_2}$$

(C.10)

Equation (C.10) is the solution for the two-PRF case, where

$\tau_1 = 21 \ \mu s$
$\tau_2 = 19 \ \mu s$
$f_1 = 47619 \text{ pulses/s}$
$f_2 = 52631 \text{ pulses/s}$

The term U will assume integral values increasing from zero until T_0 becomes positive. Note that in (C.10), $\tau_1 - \tau_2 \neq 1$.

C.2.1.2 Interpulse Intervals are Consecutive Integers

We will now consider the case where the interpulse intervals are relatively prime consecutive integers.

Let

$\tau_1 = 21 \ \mu s$
$\tau_2 = 20 \ \mu s$

These periods correspond to PRFs of

$f_1 = 47619 \text{ pulses/s}$
$f_2 = 50000 \text{ pulses/s}$

Let

$$\tau_0 = 21 \cdot 20 = 420$$

Then,

$$\frac{\tau_0}{\tau_1} = 20 \text{ μs}$$

and

$$\frac{\tau_0}{\tau_2} = 21 \text{ μs}$$

$$X_1 (20) \equiv 1 \pmod{21}$$

which is equivalent to

$$20 X_1 - 21U = 1$$

By inspection, we obtain

$$X_1 = -1, +20, +41 \text{ } et \text{ } cetera; \quad U = -1$$

Similarly, for X_2,

$$X_2 \frac{\tau_0}{\tau_2} \equiv 1 \text{ } [\text{mod}(\tau_2)]$$

$$21 X_2 \equiv 1 \pmod{20}$$

$$21 X_2 - 20U = 1$$

By inspection, we obtain
$$X_2 = +1, +21, +41 \text{ } et \text{ } cetera; \quad U = +1$$

By using the Chinese remainder theorem,

$$T_0 \equiv (-1)(20) \text{ } t_1 + (+1)(21) \text{ } t_2 \pmod{\tau_1 \tau_2}$$
$$\equiv \tau_1 t_2 - \tau_2 t_1 \pmod{\tau_1 \tau_2} \tag{C.11}$$

By the nature of congruences, (C.11) is the equivalent of

$$T_0 = \tau_1 t_2 - \tau_2 t_1 + U(\tau_1 \tau_2)$$

This can be rewritten as

$$T_0 = \frac{1}{f_1} t_2 \frac{-1}{f_2} t_1 + U \frac{1}{f_1 f_2} \tag{C.12}$$

$$T_0 = \frac{f_2 t_2 - f_1 t_1 + U}{f_1 \cdot f_2} \tag{C.13}$$

$$\tau_1 \tau_2 \left(\frac{1}{\tau_1} - \frac{1}{\tau_2} \right) = \tau_2 - \tau_1$$

$$\frac{1}{f_1} \cdot \frac{1}{f_2} (f_1 - f_2) = \tau_2 - \tau_1$$

Let

$$\tau_1 - \tau_2 = 1 \text{ (consecutive integers)}$$

Then,

$$f_1 f_2 = f_2 - f_1 \tag{C.14}$$

Substituting (C.14) into (C.13), we have

$$T_0 = \frac{f_2 t_2 - f_1 t_1 + U}{f_2 - f_1} \tag{C.15}$$

Equations (C.13) and (C.15) are equivalent, and apply to the two-PRF situation in the special case where the relatively prime interpulse intervals differ from each other by one unit. The term U has an integral value increasing from zero to the lowest integer for which equations (C.13) and (C.15) are positive.

C.2.2 Effect of Multiple Targets

In the preceding discussion, we dealt with determining the range to a single target. Although unlikely, under certain circumstances, multiple true targets might be present within the detection circuit of the radar. This possibility is discussed in

Section C.1. We will now make an analysis to determine in a quantitative manner the displacement of the ghost targets with respect to the true targets.

If more than one target is present, the range measurement becomes more complicated than with a single target because the radar cannot associate a target at one PRF with the same target at the other PRF. The radar will therefore compute the correct ranges, but will also compute ranges to ghost targets which do not actually exist.

Of interest is to compare (C.7) and (C.11). Equation (C.11) can be rewritten (using the positive value of $X_1 = 20$) as

$$T_0 \equiv (20)(20)\ t_1 + (1)(21)\ t_2\ (\text{mod}\ \tau_1\tau_2)$$
$$T_0 \equiv 400\ t_1 + 21\ t_2\ (\text{mod}\ 420) \tag{C.16}$$

This solution to T_0 for the case of consecutive integer values of τ_1 and τ_2 is analogous to the solution given in (C.7) for the nonconsecutive integer values. Both are of the form:

$$T_0 \equiv G\ t_1 + H\ t_2\ (\text{mod}\ \tau_1\tau_2)$$

Therefore, the creation of ghosts is a similar problem in each case. Figure C.2 shows the timing situation for the two-PRF case, with two targets present. The subscript j associated with t_{0j} and t_{jk} refers to the number of targets. The subscript k associated with t_{jk} refers to the number of the PRF. Correct measurement of T_{0j} is made when t_{11} and t_{12} are used for the calculation of T_{01}, and when t_{21} and t_{22} are used for calculating T_{02}. Unfortunately, all the targets appear the same to the radar, and its circuits will make measurements by combining each time interval in PRF 1 with each time interval in PRF 2. Equation (C.16) expresses the true range in question when the true values are used for t_{j1} and t_{j2}. As previously discussed, however, the radar cannot be sure of using the correct values. As shown in Figure C.2,

$$t_{21} = t_{11} + \Delta R$$
$$t_{22} = t_{12} + \Delta R$$

The four possible combinations for computed T_0 are

(a) $T_0 \equiv G(t_{11}) + H(t_{12})\ (\text{mod}\ \tau_1\tau_2)$
(b) $T_0 \equiv G(t_{11} + \Delta R) + H(t_{12} + \Delta R)\ (\text{mod}\ \tau_1\tau_2)$
(c) $T_0 \equiv G(t_{11} + \Delta R) + H(t_{12})\ (\text{mod}\ \tau_1\tau_2)$
(d) $T_0 \equiv G(t_{11}) + H(t_{12} + \Delta R)\ (\text{mod}\ \tau_1\tau_2)$

Of these, (a) represents the true range of target 1, and (b) represents the true range of target 2. The last two combinations (c) and (d) represent ghost targets produced by the radar operating in the presence of multiple targets.

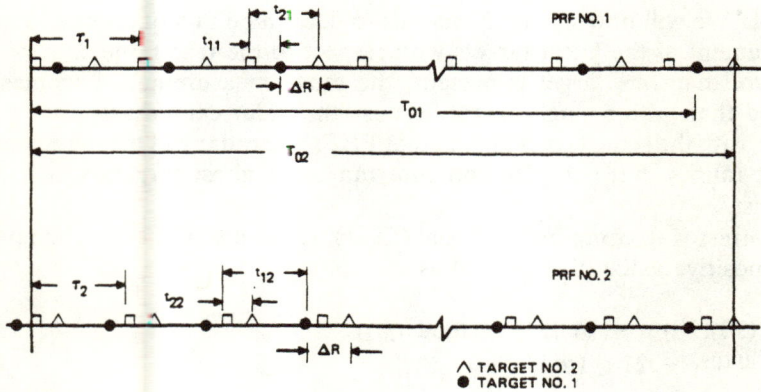

Figure C.2 Two-PRF case, two targets.

In these equations, from (C.16),

$$G = 400$$
$$H = 21$$
$$\tau_1 \tau_2 = 420$$

This is the situation where

$$PRF\ 1 = 47,619 \text{ pulses/s}$$
$$PRF\ 2 = 50,000 \text{ pulses/s}$$

With these PRFs, the interpulse spacing at PRF 1 = 1/47,619 = 21 μs and the interpulse spacing at PRF 2 = 1/50,000 = 20 μs. If the resolution interval of the radar receiver is 1 μs, there are 20 × 21 = 420 possible true ranges which can be measured by the radar. For each of the true ranges, there are a total of four possible combinations calculated by the radar, as stated previously.

The displacement of the ghost targets was determined by computer analysis, the results of which are shown in Table C.1. The computer selected typical ranges from the 420 possible ranges and, for each of these, calculated the range intervals in which the four targets (two real and two ghost) appeared. This was done for a typical spacing between the two targets of 3 μs. Consider the meaning of the results of Table C.1. The first range interval used in the calculation was 400 μs. The true target, therefore, is at 400 μs by definition. The second real target is spaced 3 μs from it, or at 403 μs. These two real targets are represented by the 2 in the 400 to 499 range interval column. The two "ghost" targets appear in the intervals 0 to 49 and 300 to 349 μs, repectively. Therefore, for the true ranges of 400 and 403 μs, the radar will compute a true target 3 μs from the real target and two "ghost" signals, one in the region 51 to 100 μs from the true target and the other in the region 351 to 400 μs from the true target. A 400 μs displacement represents a radar range displacement in the order of 30 nmi. Table C.1 shows that a similar displacement of "ghost" targets is produced for other values of true range.

The analysis in the preceding paragraphs for the two-PRF case at 50 kHz will now be extended to the 100 kHz case. The congruence which defines this case, determined in a manner similar to Equation (C.16), is

$$T_0 \equiv 100 \, t_1 + 11 \, t_2 \pmod{110} \tag{C.17}$$

Table C.1
Calculation of Targets: Two PRFs, Two Targets, 50 kHz PRF

EFFECTIVE TARGETS

Delay of Target 1 = 3 μs
PRFs = 50,000; 47,619
Target Range: *Auxiliary Target Region*

	0–49	50–99	100–149	150–199	200–249	250–299	300–349	350–399	400–449
400	1	0	0	0	0	0	1	0	2
350	0	0	0	0	0	1	0	2	1
300	0	0	0	0	1	0	2	1	0
250	0	0	0	1	0	2	1	0	0
200	0	0	1	0	2	1	0	0	0
150	0	1	0	2	1	0	0	0	0
100	1	0	2	1	0	0	0	0	0
50	0	2	1	0	0	0	0	0	1

Equation (C.17) is based on interpulse intervals that are relatively prime consecutive integers of 11 and 10 μs, respectively, corresponding to PRFs of 90,909 and 100,000 pulses/s.

The results of a computer analysis of ghost targets for this case is shown in Table C.2. The maximum unambiguous range for this condition is 110 μs because of the modular nature of (C.17). The spacing of targets, real and ghost as calculated by the radar, is shown in Table C.2. Consider the first actual target range of 100 μs with the second target displaced 3 μs. The two computed real targets will occur at 100 and 103 μs, respectively. These are represented in Table C.2 by the 2 in the 100 to 119 μs region. The two ghosts occur in the 20 to 29 and 60 to 70 μs regions, respectively. The first ghost will therefore be 61 to 80 μs from the true 100 μs target. The second ghost will be 21 to 40 μs from the true 100 μs target. These spacings, when converted to radar miles, are 5.0 to 6.6 nmi for the first ghost, and 1.7 to 3.3 nmi for the second ghost.

Table C.2
Calculation of Targets: Two PRFs, Two Targets, 100 kHz PRF

Delay of Target 1 = 3 µs
PRFs = 100000, 90909
Target Range:

	Effective Targets					
				Auxiliary Target Region		
	0–19	20–39	40–59	60–79	80–99	100–119
100	0	1	0	1	0	2
80	1	0	1	0	2	0
60	0	1	0	2	1	0
40	1	0	2	1	0	0
20	0	2	1	0	0	1

When compared with those of Table C.1, the results of Table C.2 show that the displacement of the ghost targets from the true targets is greater for lower PRFs.

C.3 THREE-PRF CASE

C.3.1 Range Calculation

The three-PRF case is an extension of the two-PRF case. The applicable time relationships are shown in Figure C.3. In this case,

$$T_0 = k\tau_1 + t_1$$
$$= l\tau_2 + t_2$$
$$= m\tau_3 + t_3 \qquad (C.18)$$

By the theory of congruences, (C.18) can be expressed as

$$T_0 \equiv t_1 \ (\text{mod } \tau_1)$$
$$\equiv t_2 \ (\text{mod } \tau_2)$$
$$\equiv t_3 \ (\text{mod } \tau_3) \qquad (C.19)$$

By using the Chinese remainder theorem, if τ_1, τ_2, and τ_3 are relatively prime by pairs,

$$T_0 \equiv (t_1) \frac{T_0}{\tau_1} (X_1) + (t_2) \frac{T_0}{\tau_2} (X_2)$$

$$+ (t_3) \frac{T_0}{\tau_3} (X_3) \ (\text{mod } \tau_1\tau_2\tau_3) \qquad (C.20)$$

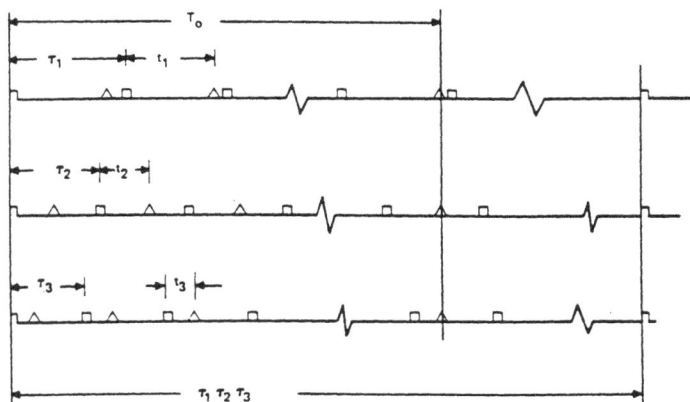

Figure C.3 Time relationships in the three-PRF case.

where

$$\tau_0 = (\tau_1)(\tau_2)(\tau_3) \tag{C.21}$$

$$\frac{\tau_0}{\tau_1} X_1 \equiv 1 \ (\text{mod } \tau_1)$$

$$\frac{\tau_0}{\tau_2} X_2 \equiv 1 \ (\text{mod } \tau_2) \tag{C.22}$$

$$\frac{\tau_0}{\tau_3} X_3 \equiv 1 \ (\text{mod } \tau_3) \tag{C.23}$$

Linear congruences in (C.21), (C.22) and (C.23) can be solved by the method of Ore [3]. Consider a typical example to illustrate the method. Let

$$\tau_1 = 13$$
$$\tau_2 = 12$$
$$\tau_3 = 11$$
$$t_1 = 4$$
$$t_2 = 6$$
$$t_3 = 9$$

where units are given in μs.

$$\tau_0 = 13 \cdot 12 \cdot 11 = 1716$$

$$\frac{\tau_0}{\tau_1} = (12)(11) = 132$$

$$\frac{\tau_0}{\tau_2} = (13)(11) = 143$$

$$\frac{\tau_0}{\tau_3} = (13)(12) = 156$$

Then,

$$(12)(11)\ X_1 \equiv 1 \ (\text{mod } 13)$$

This is equivalent to

$$(12)(11)\ X_1 - 13U = 1 \tag{C.24}$$

where U is an integer.

Rewriting, X_1 and U can be obtained as follows [4]:

In the solution of a linear congruence where the remainder is unity and the factors are relatively prime, there will always be a remainder of unity when the factoring shown in (C.25) is performed $(13 = 2.6 + unity)$. The quotients of equation (C.25) can be arranged and multiplications can be performed using the following configuration.

$$
\begin{aligned}
132X - 13U &= 1 \\
132 &= 13 \cdot 10 + 2 \\
13 &= 2 \cdot 6 + 1
\end{aligned}
\tag{C.25}
$$

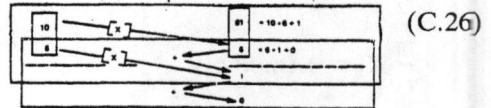
(C.26)

The 1 and 0 are arbitrarily beneath the dotted line in the configuration to initiate the sequence. After the procedure shown in configuration (C.26) has been followed, starting at the bottom of the configuration with $6 \cdot 1 + 0$, the quantitites 61 and 6 are the absolute values for U and X_1, respectively. Numerical signs can be assigned by inspection, giving the result:

$$X_1 = -6$$
$$U = -61$$

After substitution of values for X_1 and U_1, (C.24) becomes

$$132\,(-6) - 13\,(-61) = 1$$

which is obviously true.

Equation (C.24) resulted from the congruence

$$132\,X_1 \equiv 1 \text{ (mod 13)}$$

The value of -6 for X_1 satisfies the congruence. The nature of a congruence is such that integral multiples of the modulus will also satisfy the congruence. Accordingly, $(-6 + 13) = +7$, $(-6 + 13 \cdot 2) = +20$, *et cetera*, will also satisfy the congruence. Any of the values, positive or negative, which satisfy the congruence may be used in the Chinese remainder theorem equation (C.20).

Repetition of the calculation of (C.26) establishes values for X_2 and X_3 as follows:

$$X_2 = -1,\ +11,\ et\ cetera$$
$$X_3 = -5,\ +6,\ et\ cetera$$

Calculated values can be substituted in (C.20) to give

$$T_0 \equiv (4)(132)(-6) + (6)(143)(-1) + (9)(156)(-5)(\text{mod }1716)$$

Because working with positive quantities is easier than negative ones, positive values for X_1, X_2, and X_3 will be used. Now,

$$
\begin{aligned}
T_0 &\equiv (4)(132)(7) + (6)(143)(11) + (9)(156)(6)(\text{mod }1716)\\
&\equiv (4)(924) + (6)(1573) + (9)(936)(\text{mod }1716)\\
&\equiv 3696 + 9438 + 8424 \text{ (mod }1716)\\
&\equiv 21{,}558 \text{ (mod }1716) \tag{C.27}
\end{aligned}
$$

The modular nature of (C.27) means that solutions to it will repeat at intervals of the modulus (i.e., 1716, in this case). Therefore, if the computed value of T_0 exceeds 1716, enough integral multiples of 1716 must be substracted to give a value less than 1716. Accordingly, in (C.27), the desired value for T_0 is

$$T_0 = 21{,}558 - (12)(1716) = 966 \text{ μs} \tag{C.28}$$

The theory of congruences shows that the solution to a congruence repeats after each integral value of the modulus. As applied to the problem of measurement of true range in a radar, this means that the maximum unambiguous range will be equal to the product of the individual interpulse spacings expressed in radar range

(12.19 μs = 1 nmi). For the case illustrated, the maximum unambiguous range will be 1716 μs = 141 nmi.

C.3.2 Effect of Multiple Targets

The preceding discussion dealt with the determination of range to a single target. As discussed in Section C.1, under certain circumstances, multiple targets might appear simultaneously within the detecting circuit of the pulsed doppler radar. In this section, we will consider the displacements of the ghost targets produced by three true targets that are simultaneously present.

Consider the situation shown in Figure C.4, where three targets are shown and each is defined by an individual symbol. If the radar received targets with associated symbols, as in Figure C.4, it could perform measurements on the individual targets at each PRF and determine the three true ranges without ambiguity. Unfortunately, all returns look alike to the radar, as shown in Figure C.5. Figure

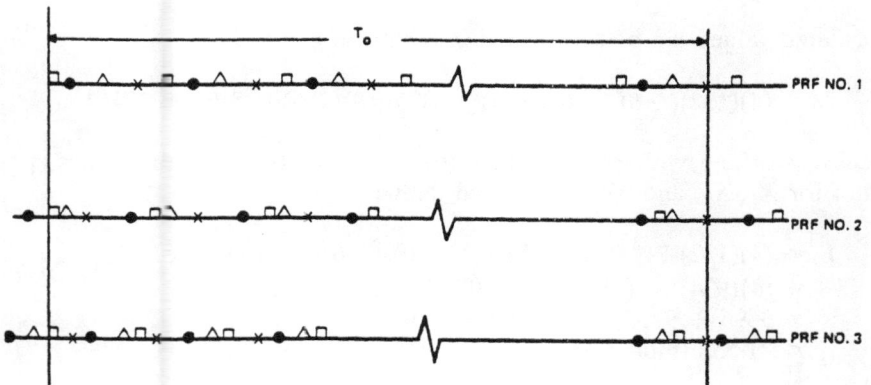

Figure C.4 Time relationship: three PRFs, three targets.

C.5 is redrawn in Figure C.6 with a number assigned to each return. The radar should perform measurements on 1, 4, 7 to determine range to target 1; on 2, 5, 8 to determine range to target 2; and on 3, 6, 9 to determine range to target 3. As mentioned above, the radar cannot make this distinction. So, for each return on PRF 1, there are three returns of PRF 2 and three returns on PRF 3 for the radar to compare. There are, therefore, $3 \cdot 3 \cdot 3 = 27$ possible ranges which the radar will compute. Of these, only three are correct. If, instead of three targets, there were four, the number of tagets computed would be $4 \cdot 4 \cdot 4 = 64$, of which only four would be correct. In general, if there are k targets, and n PRFs, the total

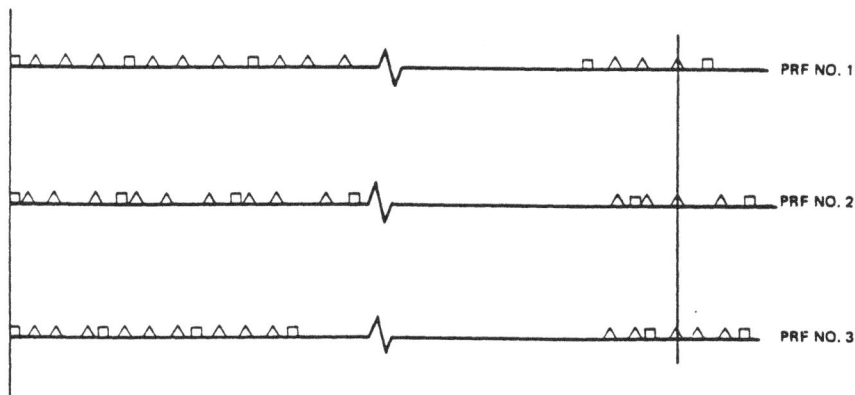

Figure C.5 Time relationship: targets unidentified.

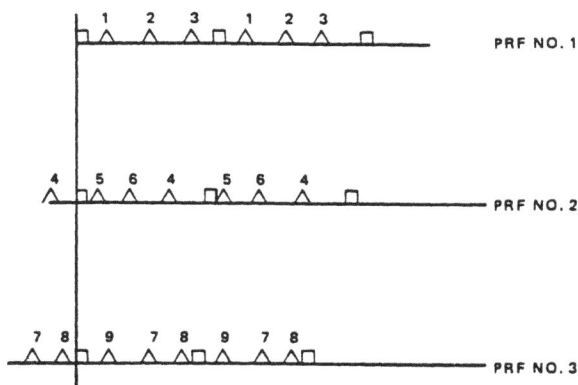

Figure C.6 Target combinations.

number of targets as seen by the radar is $(k)^n$, and the number of ghost targets is $(k)^n - k$.

As previously discussed, for the situation in which three targets are present, $(k)^n - k$ ghost targets are possible. This analysis will assume a radar that employs three PRFs; 27 targets will be computed by the radar. Of these, three are actual targets and 24 are ghost targets.

We will now make a quantitative analysis of the amount of delay produced by each of the ghost targets. A timing diagram reflecting the position of the targets is shown in Figure C.7. Distinctive symbols are assigned to each of the three targets.

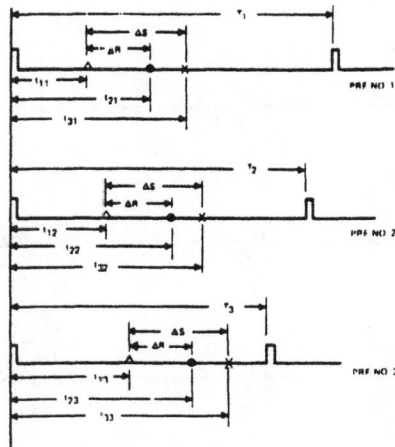

Figure C.7 Time Relationship: three PRFs, three targets.

As shown, the spacing between the first target and each of the other two is the same, regardless of which PRF is receiving the signals. The difference between the situations (as received by each of the PRFs) is the time between the transmitted pulse and the first of the three target pulses. This difference is shown as t_{11}, t_{12}, and t_{13}.

The first target is delayed ΔR μs from the true target; the second is delayed ΔS μs. Therefore, as shown in Figure C.7,

$$t_{21} = t_{11} + \Delta R$$
$$t_{31} = t_{11} + \Delta S$$
$$t_{22} = t_{12} + \Delta R$$
$$t_{32} = t_{12} + \Delta S$$
$$t_{23} = t_{13} + \Delta R$$
$$t_{33} = t_{13} + \Delta S$$

We will assume that the first radar chosen for analysis has a nominal PRF of 100 kHz. The interpulse interval is therefore 10 μs at this PRF. So that the interpulse intervals may be consecutive and relatively prime, intervals of 11, 12, and 13 μs will be used. These are equivalent to PRFs of 90,909, 83,333, and 76,923 pulses/s respectively. These PRFs are the same as those used in the calculation of (C.27), so the congruence applicable to this situation, as in the case of the example of (C.27), is

$$T_0 \equiv 924t_1 + 1573t_2 + 936t_3 \;(\text{mod } 1716) \tag{C.29}$$

Equation (C.29) therefore gives the true range to the target. The range to each of the 27 computed targets can be written as

$$T_0 = 924t_{x1} + 1573t_{y2} + 936t_{z3} \text{ (mod 1716)} \tag{C.30}$$

where x, y, and z can assume values from 1 to 3, representing the three targets.

For the situation defined by (C.30) in which the radar PRF is nominally 100 kHz, there are 1716 true target ranges possible if the resolution of the radar is 1 μs. With three targets simultaneously present, three of the twenty-seven targets which the radar computes will be the actual targets and the remaining twenty-four will be ghost targets. A computer was used to determine the displacement of the ghost targets as compared to the true target. It computed the twenty-seven real and "ghost" targets associated with each of the 1716 possible true ranges. The second and third targets are arbitrarily spaced 3 and 4 μs, respectively, from the first target. We arbitrarily decided that any target of range greater than 1000 μs would be unrealistic; 1000 μs represents approximately 80 nmi, and the radar is unlikely to be interested in true targets at greater ranges. This is strictly an arbitrary limitation, and may be made more or less restrictive.

The results of this analysis of the spread in ghost target positions are given in Table C.3, for the three-target, three-PRF, 100 kHz case. The printout is analogous to that of Table C.2 for the two-target, two-PRF, 100 kHz case. In Table

Table C.3
Calculation of Targets: Three PRF's, Three Targets,
100 kHz PRF

Effective Targets
Delay of Target No. 1 = 3 μs
Delay of Target No. 2 = 4 μs

PRFs = 90909, 83333, 76923

Target Range:	Auxiliary Target Region									
	0–99	100–199	200–299	300–399	400–499	500–599	600–699	700–799	800–899	900–999
900	0	2	2	1	1	3	1	3	0	3
800	2	2	1	1	3	1	3	0	3	1
700	2	1	1	3	1	3	0	3	1	1
600	1	1	3	1	3	0	3	1	1	1
500	1	3	1	3	0	3	1	1	1	2
400	3	1	3	0	3	1	1	1	2	2
300	1	3	0	3	1	1	1	2	2	2
200	3	0	3	1	1	1	2	2	2	2
100	0	3	1	1	1	2	2	2	2	0

C.3, the results for only a limited number of selected ranges to the first target (at 100 μs intervals) are printed to limit the size of the printout. Examination of Table C.3 shows that the number of targets computed by the radar (real targets as well as ghosts) varies depending on the true range, but in general is approximately 15. Of these, three are actual targets and the remainder are ghosts. Consider now the spread of ranges of the "ghost" targets. The range to the first true target in Table C.3, is 900 μs. The other two targets are spaced at 3 and 4 μs, respectively, from the first target; therefore, the other two true targets are at ranges of 903 and 904 μs. As shown in Table C.3, there are these three targets in the 900 to 999 μs region. There are, in addition, thirteen ghost targets spaced as follows:

Region (μs)	Number in Region
700–799	3
600–699	1
500–599	3
400–499	1
300–399	1
200–299	2
100–199	2

The real targets are, by definition, at ranges of 900, 903 and 904 μs. Therefore, the ghosts are apparently spaced 100 μs and greater from the real target. The minimum spacing is therefore on the order of 8 nmi. The maximum spacing is on the order of 66 nmi:

$$\frac{900 - 100 \text{ μs}}{12.19 \text{ μs/nmi}} = 66 \text{ nmi}$$

Note that the choice of 3 and 4 μs as spacings in this analysis was entirely arbitrary.

Our analysis concerned a radar with a basic PRF that was 100 kHz. A radar with a basic PRF of 50 kHz will now be considered. The interpulse interval for a PRF of 50 kHz is $1/50,000 = 20$ μs. So that the interpulse intervals be relatively prime consecutive integers, the interpulse intervals are assumed to be 19, 20, and 21 μs, corresponding to PRFs of 52,631, 50,000, and 47,619 pulses/s, respectively. The true range is given by the congruence:

$$T_0 \equiv 4180t_1 + 7581t_2 + 4200t_3 \pmod{7980} \tag{C.31}$$

This congruence was determined in the same manner as for (C.29).

The results are given in Table C.4. As in the case of 100 kHz PRF, the ghost targets are spaced many miles from the true targets. There are, however, fewer ghost targets produced because the maximum unambiguous range (7980 μs) in the

Table C.4
Calculation of Targets: Three PRFs, Three Targets,
50 kHz PRF

Effective Targets
Delay of Target No. 1 = 3 μs
Delay of Target No. 2 = 4 μs

PRFs = 52,631; 50,000; 47,619

Target Range:	Auxiliary Target Region									
	0–99	100–199	200–299	300–399	400–499	500–599	600–699	700–799	800–899	900–999
900	1	1	0	0	1	2	0	0	0	3
800	1	0	0	1	2	0	0	0	3	0
700	0	0	1	2	0	0	0	3	0	0
600	0	1	2	0	0	0	3	0	0	0
500	1	2	0	0	0	3	0	0	0	1
400	2	0	0	0	3	0	0	0	1	0
300	0	0	0	3	0	0	0	1	0	0
200	0	0	3	0	0	0	1	0	0	1
100	0	3	0	0	0	1	0	0	1	1

50 kHz situation is greater than that for the 100 kHz situation (1716 μs). So, in the 50 kHz situation, more of the ghost targets occur outside the arbitrary 1000 μs acceptance limit.

REFERENCES

1.   O. Ore, *Number Theory and Its History*. McGraw-Hill, New York, 1948, pp. 245–248.
2.   *Ibid*. Chapter 7.
3.   *Ibid*.
4.   *Ibid*.

Bibliography

Chapter 1:

Barton, D.K., ed., *Radars,* Seven Volume Series. Dedham, MA: Artech House, 1974–1979.
Berkowitz, R.S., ed., *Modern Radar.* New York: John Wiley and Sons, 1966.
Boyd, J.A., *et al., Electronic Countermeasures.* Los Altos, CA: Peninsula Publishing, 1978.
Brookner, E., ed., *Radar Technology.* Norwood, MA: Artech House, 1977.
Constant, J., *Defense Radar Systems.* New York: Spartan Books, 1972.
Maksimov, M.V., *et al., Radar Anti-Jamming Techniques.* Norwood, MA: Artech House, 1979.
Povejsil, D.J., R.S. Raven, and P. Waterman, *Airborne Radar.* Cambridge, MA: Boston Technical
    Publishers, 1965.
Schleher, D.C., *Introduction to Electronic Warfare.* Norwood, MA: Artech House, 1986.
Skolnik, M.I., *Introduction to Radar Systems,* New York: McGraw-Hill, 1980.
Van Brunt, L.B., *Applied ECM.* Dunn Loring, VA: EW Engineering, Inc., 1978.

Chapter 2:

Barton, D.K., *Modern Radar System Analysis.* Dedham, MA: Artech House, 1988.
Blake, L.V., *Radar Range-Performance, Analysis.* Norwood, MA: Artech House, 1986.
Boyd, J.A., *et al., Electronic Countermeasures.* Los Altos, CA: Peninsula Publishing, 1978.
Povejsil, D.J., R.S. Raven, and P. Waterman, *Airborne Radar.* Cambridge, MA: Boston Technical
    Publishers, 1965.
Ruck, G.T., ed., *Radar Cross Section Handbook.* New York: Plenum Press, 1970.
Schleher, D.C., *Introduction to Electronic Warfare.* Norwood, MA: Artech House, 1986.
Skolnik, M.I., *Introduction to Radar Systems.* New York: McGraw-Hill, 1980.
Van Brunt, L.B., *Applied ECM.* Dunn Loring, VA: EW Engineering, Inc., 1978.

Chapter 3:

Barton, D.K., ed., *Radars, Volume Seven: CW and Doppler Radars.* Dedham, MA: Artech House,
    1979.
Boyd, J.A., *et al., Electronic Countermeasures.* Los Altos, CA: Peninsula Publishing, 1978.
Evans, G.E., and H.E. Schrank, "Low Sidelobe Radar Antennas," *Microwave Journal,* July 1983,
    p. 109.

Hovanessian, S.A., "Medium PRF Performance Analysis," *IEEE Trans. on Aerospace and Electronic Systems,* May 1982, p. 286.

Hovanessian, S.A., *Radar System Design and Analysis.* Norwood, MA: Artech House, 1984.

Hovanessian, S.A., "Target Image Frequency Spectrum in Doppler Radars," *IEEE Trans. on Aerospace and Electronic Systems,* July 1974, p. 497.

Johnston, S.L., *Radar Electronic Counter-Countermeasures.* Dedham, MA: Artech House, 1979.

Maksimov, M.V., *et al. Radar Anti-Jamming Techniques.* Norwood, MA: Artech House, 1979.

Povejsil, D.J., R.S. Raven, and P. Waterman, *Airborne Radar.* Cambridge, MA: Boston Technical Publishers, 1965.

Schleher, D.C., *Introduction to Electronic Warfare.* Norwood, MA: Artech House, 1986.

Skolnik, M.I., *Introduction to Radar Systems.* New York: McGraw-Hill, 1980.

Van Brunt, L.B., *Applied ECM.* Dunn Loring, VA: EW Engineering, 1978.

Chapter 4:

Barton, D.K., ed., *Radars, Volume Seven: CW and Doppler Radars.* Dedham, MA: Artech House, 1979.

Boyd, J.A., *et al., Electronic Countermeasures.* Los Altos, CA: Peninsula Publishing, 1978.

Evans, G.E., and H.E. Schrank, "Low Sidelobe Radar Antennas," *Microwave Journal,* July 1983, p. 109.

DiFranco, J.V., and W.L. Rubin, *Radar Detection.* Norwood MA: Artech House, 1980.

Hovanessian, S.A., *Radar System Design and Analysis.* Norwood, MA: Artech House, 1984.

Hovanessian, S.A., "Signal to Noise Ratio Calculations in Pulse and Pulse Doppler Radars," *IEEE Trans. on Aerospace and Electronic Systems,* September 1981, p. 722.

Johnston, S.L., ed., *Radar Electronic Counter-Countermeasures.* Dedham, MA: Artech House, 1979.

Maksimov, M.V., *et al. Radar Anti-Jamming Techniques.* Norwood, MA: Artech House, 1979.

Nathanson, F.E., *Radar Design Principles.* New York: McGraw-Hill, 1969.

Oppenheim, A.V., and R.W. Schafer. *Digital Signal Processing.* Englewood Cliffs, NJ: Prentice-Hall, 1975.

Povejsil, D.J., R.S. Raven, and P. Waterman, *Airborne Radar.* Cambridge, MA: Boston Technical Publishers, 1965

Schleher, D.C., *Introduction to Electronic Warfare.* Norwood, MA: Artech House, 1986.

Van Brunt, L.B., *Applied ECM.* Dunn Loring, VA: EW Engineering, 1978.

Chapter 5:

Barton, D.K., ed., *Radars, Volume Seven: CW and Doppler Radars.* Dedham, MA: Artech House, 1979.

Blake, L.V., *Radar Range-Performance Analysis.* Norwood, MA, Artech House, 1986.

Boyd, J.A., *et al., Electronic Countermeasures.* Los Altos, CA: Peninsula Publishing, 1978.

Hovanessian, S.A., "Medium PRF Performance Analysis," *IEEE Trans. on Aerospace and Electronic Systems,* May 1982, p. 286.

Hovanessian, S.A., *Radar System Design and Analysis.* Norwood, MA: Artech House, 1984.

Hovanessian, S.A., "Signal to Noise Ratio Calculations in Pulse and Pulse Doppler Radars," *IEEE Trans. on Aerospace and Electronic Systems,* September 1981, p. 722.

Hovanessian, S.A., "Target Image Frequency Spectrum in Doppler Radars," *IEEE Trans. on Aerospace and Electronic Systems,* July 1974, p. 497.

Maksimov, M.V., *et al., Radar Anti-Jamming Techniques.* Norwood, MA: Artech House, 1979.

Povejsil, D.J., R.S. Raven, and P. Waterman, *Airborne Radar.* Cambridge, MA: Boston Technical Publishers, 1965.
Rangel, M.B., D.H. Mooney, and W.H. Long, "F-16 Pulse Doppler Radar (AN/APG—66) Performance," *IEEE Trans. on Aerospace and Electronic Systems,* January 1983, p. 147.
Skolnik, M.I., *Introduction to Radar Systems.* New York: McGraw-Hill, 1980.
Van Brunt, L.B., *Applied ECM.* Dunn Loring, VA: EW Engineering, 1978.

Chapter 6:

Barton, D.K., ed., *Radars, Volume Seven: CW and Doppler Radars.* Dedham, MA: Artech House, 1979.
Barton, D.K., ed., *Radars, Volume Four: Radar Resolution and Multipath Effects.* Dedham, MA: Artech House, 1975.
Boyd, J.A., *et. al., Electronic Countermeasures.* Los Altos, CA: Peninsula Publishing, 1978.
Brookner, E., ed., *Radar Technology.* Dedham, MA: Artech House, 1977.
Evans, G.E., and H.E. Schrank, "Low Sidelobe Radar Antennas," *Microwave Journal,* July 1983, p. 109.
Hovanessian, S.A., *Radar System Design and Analysis.* Norwood, MA: Artech House, 1984.
Hovanessian, S.A., "Signal to Noise Ratio Calculations in Pulse and Pulse Doppler Radars," *IEEE Trans. on Aerospace and Electronic Systems,* September 1981, p. 722.
Kahrilas, P.J., *Electronic Scanning Radar System.* Dedham, MA: Artech House, 1976.
Lewis, B.L., F.F. Kretschmer Jr., and W. Shelton, *Aspects of Radar Signal Processing.* Norwood, MA: Artech House, 1986.
Maksimov, M.V., *et al., Radar Anti-Jamming Techniques.* Norwood, MA: Artech House, 1979.
Oppenheim, A.V., and R.W. Schafer, *Digital Signal Processing.* Englewood Cliffs, NJ: Prentice-Hall, 1975.
Povejsil, D.J., R.S. Raven, and P. Waterman, *Airborne Radar.* Cambridge, MA: Boston Technical Publishers, 1965.
Schleher, D.C., *Introduction to Electronic Warfare.* Norwood, MA: Artech House, 1986.
Skolnik, M.I., *Introduction to Radar Systems.* New York: McGraw-Hill, 1980.

Chapter 7:

Boyd, J.A., *et al., Electronic Countermeasures.* Los Altos, CA: Peninsula Publishing, 1978.
Hovanessian, S.A., *Introduction to Synthetic Array and Imaging Radars.* Norwood, MA: Artech House, 1980.
Hovanessian, S.A., *Radar System Design and Analysis.* Norwood, MA: Artech House, 1984.
Hovanessian, S.A. and J.C. Naviaux, "Tactical Uses of Imaging Radars," *Microwave Journal,* February 1984, p. 109.
Kovaly, J.J., *Synthetic Aperture Radar.* Dedham, MA: Artech House, 1976.
Oliner, A.A., and G.H. Knittel, eds., *Phased Array Antennas.* Dedham, MA: Artech House, 1970.
Oppenheim, A.V., and R.W. Schafer, *Digital Signal Processing.* Englewood Cliffs, NJ: Prentice-Hall, 1975.
Richaczek, A.W., *Principles of High Resolution Radar.* New York: McGraw-Hill, 1969.
Schleher, D.C., *Introduction to Electronic Warfare.* Norwood, MA: Artech House, 1986.
Skolnik, M.I., *Introduction to Radar Systems.* New York: McGraw-Hill, 1980.

Chapter 8:

Boyd, J.A., *et al.*, *Electronic Countermeasures*. Los Altos, CA: Peninsula Publishing, 1978.
Oppenheim, A.V., and R.W. Schafer, *Digital Signal Processing*. Englewood Cliffs, NJ: Prentice-Hall, 1975.
Schleher, D.C., *Introduction to Electronic Warfare*. Norwood, MA: Artech House, 1986.
Van Brunt, L.B., *Applied ECM*. Dunn Loring, VA: EW Engineering, 1978.

Chapter 9:

Boyd, J.A., *et al.*, *Electronic Countermeasures*. Los Altos, CA: Peninsula Publishing, 1978.
Brookner, E., ed., *Radar Technology*. Norwood, MA: Artech House, 1977.
Evans, G.E., and H.E. Schrank, "Low Sidelobe Radar Antennas," *Microwave Journal*, July 1983, p. 109.
Ewell, G.W., *Radar Transmitters*. New York: McGraw-Hill, 1981.
DiFranco, J.V., and W.L. Rubin, *Radar Detection*. Norwood, MA: Artech House, 1980.
Hovanessian, S.A., *Introduction to Synthetic Array and Imaging Radars*. Norwood, MA: Artech House, 1980.
Hovanessian, S.A., *Radar System Design and Analysis*. Norwood, MA: Artech House, 1984.
Johnston, S.L., ed., *Millimeter Wave Radar*. Dedham, MA: Artech House, 1980.
Johnston, S.L., ed., *Radar Electronic Counter-Countermeasures*. Dedham, MA: Artech House, 1979.
Kovaly, J.J., *Synthetic Aperture Radar*. Dedham, MA: Artech House, 1976.
Lewis, B.L., F.F. Kretschmer, Jr., and W. Shelton, *Aspects of Radar Signal Processing*. Norwood, MA: Artech House, 1986.
Maksimov, M.V., *et al.*, *Radar Anti-Jamming Techniques*. Norwood, MA: Artech House, 1979.
Seashore, C.R., "MM Wave Sensors for Missile Guidance," *Microwave Journal*, September 1983, p 133.

General

Barton, D.K., and H.R. Ward, *Handbook of Radar Measurements*. New Jersey: Prentice Hall, 1969.
Blake, L.V., *Antennas*. Norwood, MA: Artech House, 1984.
Cooley, J.W. and J.W. Tukey, "An Algorithm for the Machine Calculation of the Complex Fourier Series," *Mathematics of Computation*, April 1965, p. 297.
Hieslmair, H., C. De Santis, and J.J. Wilson, "State of the Art of Solid State and Tube Transmitters," *Microwave Journal*, October 1973, p. 46.
Law, P.E., Jr., *Shipboard Antennas*. Norwood, MA: Artech House, 1983.
Long, M.W., *Radar Reflectivity of Land and Sea*. Norwood, MA: Artech House, 1983.
Meeks, M.L., *Radar Propagation at Low Altitudes*. Norwood, MA: Artech House, 1982.
Ruck, G.T., ed., *Radar Cross Section Handbook*. New York: Plenum Press, 1970.
Skolnik, I.M., ed., *Radar Handbook*. New York: McGraw-Hill, 1970.
Stevens, M.C., *Secondary Surveillance Radar*. Norwood, MA: Artech House, 1988.
Toomay, J.C., *Radar Principles for Non-Specialists*. Belmont, CA: Lifetime Learning Publications, 1982.

Index

Multiple search radar cross-polarization ECM, 105
Multiple targets, 76
  range confusion, 85
  tracking, 110
  in velocity, 79
  in velocity for pulsed doppler radar, 77

Noise jammer time-sharing, 181
  delay, 72
  effectiveness, 75
Noise system, 41
Noncoherent microwave signal storage, 128
Noncoherent radar vulnerability, 85
Nutating antenna, 135
Nutation, 135
  crossover, 136
  modulation, 135–136

Passive countermeasures, 3, 8, 191–192
Phase front shift, 150
Phase modulated pulse compression, 46–47
Phase monopulse ECM, 150
  error signal, 146–147
  radar, 146
Phased array antenna, 172–173
  focusing, 123
Power dilution
  in frequency, 177
  in space, 178
  in time, 178
Power reflected, 24
PPI scope, 55–56
PRF agile radars, 82
PRF agility, 113
PRF jitter, 60
PRF limit, 58
Pseudorandom noise, 181
Pull-off program (RGPO), 116
Pulse compression, 44–47, 196–197
Pulse descriptor words, 186
Pulsed doppler ECM system block diagram, 77–78
Pulsed doppler radar, 63, 65
  ECM, 77–78
  range ambiguity, 67–68
  range tracking, 120–121
  vulnerability, 71
  range track ECM, 121–122
Pulse radar, 57
  range measurement, 53

range track ECM, 113
range tracking, 111
Pulse repetition rate limit, 58

Radar challenges, 191
Radar concepts, 12
Radar developments, 192–195
Radar displays, 54–56
Radar environment, 8
Radar equation, 26, 28
  example, 29
  graph, 30
Radar technology, 196
Radar tracking gate, 113
Range ambiguity resolver, 79
Range equation, 52
Range gate pull-off (RGPO), 114–117
  direction, 116
  J/S, 115
  program, 116
  recycling, 118–119
Range gate tracking, 111
Range gates, 111
Range guard gate, 129
Range measurement, 12–13
Range rate measurement, 14
Range track loop, 111
  acceleration limit, 117–118
  bandwidth, 121
  limits, 116
Range *versus* velocity ambiguity, 68
Range *versus* angle display, 56
Receiver frequency requirement, 184
Receiver requirements, 4
Reflected signal power, 22–23
Repeater system, 38
Resource management receiver requirements, 183–184
Resource sharing
  in frequency, 179–180
  in polarization, 183
  in space, 182–183
  in time, 180–181
RGPO and VGPO compatibility, 131
RGPO *versus* interceptor radars, 167

S equation, 23–29
Search radar
  angle countermeasures, 87
  angle ECM, 107
  angle measurement, 87

www.ingramcontent.com/pod-product-compliance
Lightning Source LLC
Chambersburg PA
CBHW021429180326
41458CB00001B/197